Tracks to Innovation

Springer
New York
Berlin
Heidelberg
Barcelona
Budapest
Hong Kong
London
Milan
Paris
Santa Clara
Singapore
Tokyo

ROBERT L. FLEISCHER

Tracks to Innovation

Nuclear Tracks in Science and Technology

With 92 Figures

Springer

Robert L. Fleischer
Department of Geology
Union College
Schenectady, NY 12308-3146
USA

Library of Congress Cataloging-in-Publication Data
Fleischer, R.L. (Robert Louis), 1930-
Tracks to innovation : nuclear tracks in science and technology / Robert L. Fleischer.
p. cm.
Includes bibliographical references and indexes.
ISBN-13: 978-1-4612-8775-9
1. Particle tracks (Nuclear physics) 2. Particle track etching. I. Title.
QC793.3.T67F54 1998
539.7'7—dc21 97-36140

Printed on acid-free paper.

Production managed by Anthony Battle; manufacturing supervised by Jeffrey Taub.
Photocomposed copy prepared from Microsoft Word files supplied by the author.
Printed and bound by Maple-Vail Book Manufacturing Group, York, PA.

9 8 7 6 5 4 3 2 1

ISBN-13: 978-1-4612-8775-9 e-ISBN-13: 978-1-4612-4452-3 SPIN 10645030
DOI: 10.1007/978-1-4612-4452-3

Dedication

To the many men and women of General Electric who contributed to the development of scientific and technological uses of etched nuclear particle tracks. Our society is in their debt for practical advances and intellectual enlightenment.

The answer is that a large company and a large laboratory can invest a small fraction of its funds in speculative ventures in research; these ventures promise, however tentatively, departures into entirely new businesses. Although existing needs and problems dictate the largest part of research in industry—or in the Nation, for that matter—experience has shown repeatedly that studies begun solely to grapple with fundamental questions of science can be rewarding to industrial technology, providing advances for which there was no recognized pre-existent need, however strong that "need" might become afterwards.

C. Guy Suits
Arthur M. Bueche
Director Emeritus and Director,
General Electric Research Laboratory

From the moment I picked up your book until I laid it down, I was convulsed with laughter. Some day I intend reading it.

Groucho Marx, to humorist S.J. Perelman

Preface

Why did I write this book? I hope to relay to the reader some image of the experiences of discovery in track etching by describing its diversity, and its major directions – how they were undertaken, why they were worth pursuing, where things went wrong, and where not. A common thread of many of these forays is the interplay between science and technology, and how they can complement one another in an industrial laboratory.

Who might want to read this book? Two directly concerned groups are those who use nuclear tracks and those whose interests are in the relations of science, technology, and society. Especially, I hope that anyone who likes to be aware of developments and diversity in both science and engineering will share some of my personal engrossment in the versatility of nuclear tracks and will enjoy some of the learning process that I experienced.

This book describes the marvelous simplicity and many uses of particle tracks in solids – records of radiation that can be displayed clearly by dunking the solid in a chemical solution. Although I was first introduced to etched tracks through the discoveries of P. B. Price and R. M. Walker, an earlier experiment in 1958 on LiF by D. A. Young is the first case of intentional track revelation by etching. Still earlier experiments on minerals – by H. Baumhauer in 1894, A. P. Honess in 1927, A. R. Patel and S. Tolansky in 1957 and L. C. Lovell in 1958 – showed etch figures that were from particle tracks, but not recognized as such at the time. Further, etching of other line-like defects (dislocations in crystals) had been done for grown-in dislocations in silicon by W. C. Dash in 1955 and for fresh dislocations (which may differ in having no localized decoration of the dislocations by impurity atoms) in LiF by J. J. Gilman and W. G. Johnston in 1956.

Young's original interests in tracks were on how radiation damages solids, and his prior work led him to regard the existence of tracks as "obvious" (correspondence, September 1994). Since at Harwell radiation damage was his interest and primary obligation, he did not pursue the curious phenomenon of track etching into related fields.

Price, Walker and I were under no such restraints. Our group at General Electric's Research Laboratory was charged with doing science – with the long-range expectation that some fraction of qualitatively new advances would be of practical use to society, and (crucially) to General Electric. And here the assumption was realized. Our work was designed as divergent research – work in which the objective was to demonstrate what could be done uniquely with tracks in solids for as many fields as might be fruitful. We three developed many thoughts – mostly for the other two to pursue. Amongst the three of us a common defense mechanism was the phrase, "It's too bad there's only so much time in the day" – sometimes without the "It's too bad."

The plan of this book is first to describe the early discoveries at General Electric and how they led to some understanding of track etching, the existence of etched tracks form, and when they can occur. The following five chapters describe advances that understanding tracks enabled in divergent areas of science and technology – noting the repeated interactions between science and its useful applications. The concluding chapter emphasizes the salutary effect of science in an industrial setting, with the interplay of science and technology enhancing innovation in both areas. The transitions from science to technology are often cited as one important reason for doing science; and many such connections in nuclear track fields are listed, as are the reverse transitions that also exist from technology to science.

The author makes no pretense of presenting in-depth studies. The emphasis here is on purposes, ideas, and results. For more thorough technical coverage up to 1974 and more details of the history of this subject, interested readers may consult our earlier book (FPW75, see below), which is listed in the concluding bibliography, along with other substantial sources in the extensive literature on nuclear tracks in solids. The references are only those that I found useful for this personal account. They are not complete, nor could they be, considering that the estimated number of papers that have been published on nuclear tracks in solids is eight thousand. The references serve as starting points for technical people who wish to delve deeper. The general reader may simply ignore them and yet emerge with an understanding of the basics of particle tracks.

Although most of the references for the experts will be cited in the text by authors(s) and date, more general references for those who want rather broadly to know more appear at the end of each chapter. These sources of information will be referred to by author initials and the year of publication, as in the preceding paragraph, where (FPW75) stands for the book by Fleischer, Price, and Walker that was published in 1975. Selected, more extensive sources are listed after Chapter 7.

Robert L. Fleischer
Union College

Acknowledgments

The author's greatest debts in the technical exploration described in this volume are to his nearest associates. Buford Price and Bob Walker began the study of tracks at General Electric and discovered track etching in mica in 1961. They invited me to join them in 1962, and we worked closely and almost continuously together until Walker left for Washington University in Saint Louis in 1966 and Price for the University of California at Berkeley in 1969. We collaborated intermittently thereafter, most extensively on Apollo projects and the 1975 monograph (FPW75). Each of them had previously achieved distinction through studies of defects on nearly perfect crystals – Walker on atomic (point) defects and Price on line defects (dislocations) in metals. And, since I had worked on the interactions of atomic defects and dislocations, we had a common background that speeded our communication and thinking. Both of them were intelligent, talented, opportunistic, and inventive. Bob was insistent on detailed planning, inclined to optimism, but with occasional morose periods ("grumpy" was his word). Buford was more uniformly enthusiastic, sometimes to the verge of impetuousness. We worked together well, and I thank them for the experience. Through them I achieved a new level of continuous activity, intellectual stimulation, and special human contacts.

In the next decade, 1970-1980, Howard R. Hart and I teamed up in a happy collaboration that took us into lunar science, cosmic rays, fission-track dating, and learning about gas flow in the earth. Another superb scientist with great technical care and insight, Howard too came into the track-detector business with a background in the physics of solids. We had earlier collaborated on studies of defects in superconductors on their current-carrying abilities.

Antonio Mogro-Campero was the only one of my four closest collaborators whose primary prior experience was with the sorts of particles that make tracks in solids. Tony came to General Electric from the world of cosmic ray physics, and we worked together mostly on the spectrum of opportunities that came from the fact that the alpha particles that are energetically released by radon leave tracks in certain polymeric detectors; this opportunity allows us to study the hazards of indoor radon, and possibilities of uranium exploration and earthquake prediction. His versatility, inventiveness, and penetrating thinking were major contributors to our work from the mid 70's to the mid 80's. Thank you Howard and Tony.

Shortly after the recognition of nuclear track-etching at General Electric, responsibility was assigned GE's Vallecitos Nuclear Laboratory at Pleasanton, California for any commercial developments that might emerge. As part of that activity, H. Ward Alter led the development of radon detectors for uranium exploration (and, later, home radon monitoring, petroleum exploration, and earthquake study) and Sidney C. Furman guided the production of filters made by track etching. We at the Research Laboratory were in their debt for their achievements, and I believe our nation is also.

I will not list here the fifty General Electric people who were co-authors of papers in this field, the more than sixty collaborators from elsewhere, nor the hundreds that made other contributions that have previously been acknowledged, but I note those numbers to show the immense debt and the thanks that I owe.

Any errors that appear are the author's responsibility, but many others deserve sincere thanks for many of the errors that the reader will not encounter – because they were uncovered with the aid of the following people: Chapter 1 was reviewed by C. A. Fleischer, E. L. Fleischer, J. D. Livingston, P. B. Price, and R. M. Walker; Chapter 2, C. P. Bean, C. A. Fleischer, and D. D. Peterson; Chapter 3, H. W. Alter and A. Mogro-Campero; Chapter 4, S. C. Bergman, C. W. Naeser, and N. D. Naeser; Chapter 5, A. Mogro-Campero and P. B. Price; Chapter 6, G. O. Arrhenius, D. Lal, and D. S. Miller; and Chapter 7, P. T. Carroll, E. L. Fleischer, H. R. Hart, Jr., J. G. Morone, R. M. Walker, and G. Wise. I am deeply indebted to M. L. Boroson and B. S. Fleischer for reviewing the entire text.

For supplying illustrations I thank H. W. Alter, C. P. Bean, M. Chapman, T. G. Knorr, D. Lal, C. R. Porter, P. B. Price, R. Spohr, L. G. Turner, NASA, and NCAR. Various people contributed other helpful comments, material, and support. I thank H. W. Alter, G. O. Arrhenius, C. P. Bean, F. Centolanzi, G. Csikai, R. A. Donelick, I. R. Duddy, R. Greaney, L. J. Lanzerotti, T. von Foerster, R. V. Wheeler, and R. C. Yoder.

I thank the following for permission to reproduce published diagrams and photographs: American Chemical Society (Figures 2-6, 2-7; 6-8, and 6-10), American Geophysical Union (Figure 1-10; 5-3, 5-4, 5-15; and 6-5), American Institute of Physics (J. Appl. Phys. and Rev. Sci. Instr., Figures 1-1, 1-2, 1-4; 2-11; and 5-11, 5-17), American Physical Society (Figures 1-5, 1-6; and 6-1, 6-2), Geochimica et Cosmochimica Acta (Elsevier Press, Figures 3-13, 4-1, and 5-1), Geophysics (Figure 3-10), Gordon and Breach (Figure 5-7), Journal of Glass Studies (Figure 4-4), Nature (Figure 4-9), Nuclear Tracks and Radiation Measurements (Figures 3-13, 3-4, 3-5, 3-8), Science (Figures 2-1, 2-2, 2-5, 2-9, 2-10, 2-11; 4-6; 5-6, 5-10; and 6-6, 6-11, 6-16), University of California Press (Figures 1-3; 2-7, 2-8, 2-13, 2-14, 2-15, 2-17; and 4-2, 4-3, 4-11), and University of Chicago Press (Figure 4-7).

Special thanks go to the late Milan D. Fiske – friend, colleague, and mentor at the General Electric Research Laboratory. Resolute advocate of science in industry, he gave me diverse, important support through many years. And to Julius Csikai of the Institute of Experimental Physics I give my profound thanks for continuing support. Finally, and most importantly, my deep thanks go to B. S. Carson of Rensselaer Polytechnic Institute, who typed and formatted this work with remarkable speed, accuracy, and infectious, helpful good spirits. She made this work possible.

Robert L. Fleischer
Union College

Contents

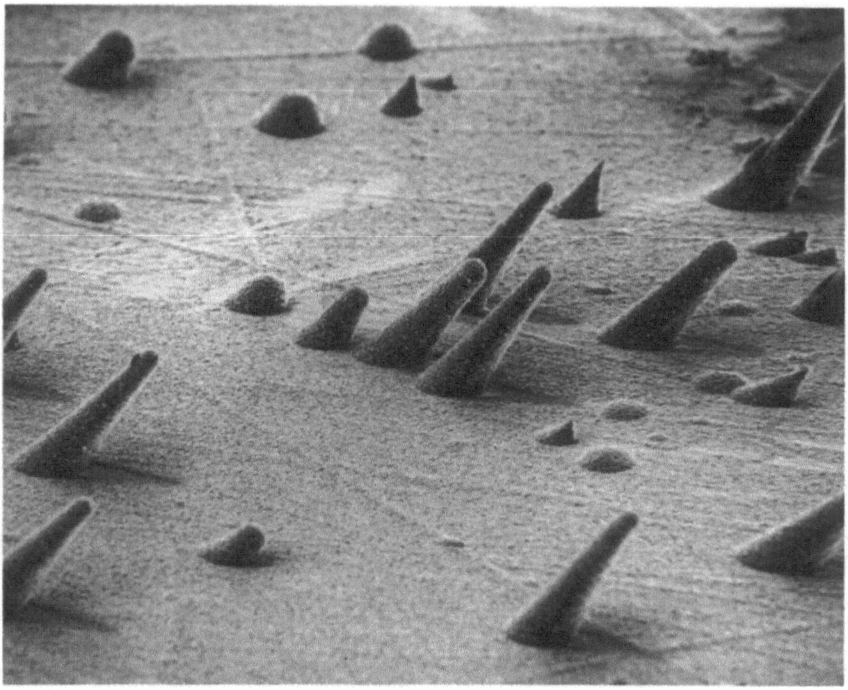

Sulfur-ion tracks from an Apollo test helmet. Replicas of the etched tracks of sulfur-32 ions were used to calibrate the helmets for cosmic ray detection. Statistical effects in how the particles slow down appear in the form of the varying length of the different tracks. The longest tracks shown are 0.8 mm. [photo by Eric Lifshin, General Electric Research and Development Laboratory, Schenectady, New York]

Chapter 1
ETCHING NUCLEAR TRACKS

... show the tracks of knowledge.
Lucretius in *On the Nature of Things*

Connections of one discovery to another are often unpredictable. Who would reason that imaginings of stored damage by cosmic particles that enter lunar rocks would lead to widespread technological applications? Yet uses of such damage range from draft beer that can be stored at room temperature to im proved oil exploration. We know now that particles can leave narrow trails of atomic disorder in solids – called *tracks*. And, conveniently, their characteristics can be displayed through the simple magic of chemical etching, which transforms tracks into holes. Many varieties of tracks exist, and the diverse solids that can record them are everywhere. For me this story began in the early 1960s.

In one of his classic spy-mysteries, *The Care of Time*, Eric Ambler seizes the reader's attention with the initial sentence, "The warning message arrived on Monday, the bomb itself on Wednesday." The analogue for our story of nuclear tracks is reversed: The bomb arrived in 1961, the letter in 1965. The first arrival was a cerebral bomb – my becoming aware of the direct observation by my future colleagues, P. Buford Price and Robert M. Walker, of fascinating, submicroscopic, radiation-damaged trails in tiny flakes of mica. The letter, by noted Berkeley physicist and isotope chemist John H. Reynolds, was the surprise of a gracious spontaneous gesture of appreciation for the scientific excitement that the three of us – Buford, Bob, and myself – had enjoyed in the intervening time, and had recently described in the journal *Science*.

The excitement stemmed from the divergent burst of uses of these esoteric-sounding tracks, which (it turns out) can be produced by nuclear particles in most solids that are electrically insulating. Much later they were found to form also in a special set of conducting solids. Price, Walker, and I were able to show first that a diverse spectrum of solids can serve as detectors for nuclear particles (fast-moving atomic nuclei) and second that with these solids many varieties of new science and technological applications become possible.

These uses of solid particle-track detectors now range from scientific uses such as measurements of ocean floor spreading at the depths of the ocean, analyzing cosmic radiation from distant parts of our galaxy, and finding remnants of radioactive plutonium that died shortly after the formation of the solar system, to technological uses, such as present-day radiation monitoring in our homes, exploration for uranium and petroleum, and attempts to predict when and where earthquakes will occur. There is a special charm to etched tracks, which at first viewing are simple but whose subtleties and reactions to their environment continue to provide new adventures and undiminished intellectual stimulation.

Perhaps nine of ten ideas in science go nowhere – often because if they are good ideas, other workers turn out to have recognized them first. And other ideas perish because our models and nature's reality just don't agree. In contrast, the recognition that tracks exist and can easily be revealed by chemical etching suddenly provided qualitatively new tools for visualizing and measuring nuclear reactions, chemical distributions in solids, and doses of neutrons and heavy nuclei. The detectors – the tools – are themselves of immense variety; documenting what substances did and did not register tracks has been interesting for its own sake and is vital to constructing models of how tracks form.

DETECTORS

Atoms are the building blocks of the world around us. Often they are caused to move rapidly, and as scientists we wish to follow their behavior through space and time. To do this we need detectors, the fundamental tools to pursue the physics of nuclear particles. This work is the story first of how the newest detectors of particle tracks – solids – allow the paths of nuclear particles to be simply revealed and second of how their uses have radiated through the sciences into diverse technologies. Shortly the earlier detectors will be described briefly and compared.

At the General Electric Research Laboratory three funny things were important to beginning the story I want to tell. Louis Navias – the guy in the lab next door – had squirreled away samples of a very stable synthetic mica created to provide good insulators that would be available in times of war, when overseas supplies might be cut off. Down the hall, John Gilman and William Johnston had achieved fame by using chemical etching to reveal crystal defects called dislocations whose motion is the basic mechanism of plastic deformation of crystalline solids. And for a commercial use, one of G.E.'s Electronics Laboratory engineers wanted a simple way to let air leak into a vacuum slowly and gently. These chance aspects of the environment at the laboratory remarkably combined to allow the discovery of submicroscopic etched tracks in mica and to reveal their simple macroscopic usefulness as nuclear particle detectors.

Discoveries can be evanescent. A modest discouragement early in a scientific exploration can render it stillborn. Yet a special advantage may encourage and continue its precarious life.

Thus, Young (1958) produced shallow etched tracks of charged particles in lithium fluoride, but saw no future in pursuing the subject. At the same laboratory Silk and Barnes (1959) used the very high magnification of a transmission electron microscope (TEM) to see such tracks directly in natural muscovite mica – but only fleetingly, since the tracks faded and were lost in the electron beam of the TEM. These workers too dropped the subject as a curiosity that was difficult to use.

Walker, as reported by Price and Walker (1962a), saw Silk and Barnes' work as an avenue to bigger game – observing cosmic ray particles in materials from the surface of the moon. Price and Walker confirmed the transient character of tracks in natural micas as viewed by an electron beam, but found that Navias' synthetic micas recorded tracks crisply and stably, as shown in Figure 1-1. Here – inside of a tiny flake of mica – lies a wondrous visual world of nuclear particles.

This minute world is transformed by simply dipping the mica into hydrofluoric acid for a few seconds (Price and Walker, 1962b). The contrast in Figure 1-2 is a negative version of the type of view in Figure 1-1. In this case thickness is used to provide contrast – so that light regions should be interpreted as thinner areas. In short, what appear as individual, well-collimated searchlights are in fact narrow (10-nm) holes that cross the flake of mica. Whether the trick of etching was inspired consciously by the technique that was used by Gilman and Johnston to study LiF is, according to Walker, likely; regardless, the work was done in an environment that was permeated with the excitement of Gilman and Johnston's work. So its subliminal presence in Price and Walker's minds may have provided the spark of invention. Etching had two powerful advantages over other methods of particle imaging: The first of these is in the permanence of etched tracks. Unlike its effect on the radiation-damaged regions that preceded etching, the electron beam does not erase holes. Now tracks could be studied at leisure. The second advantage is magnification.

Magnification was achieved by way of the third ingredient—the request for a well-defined, controlled gas leak. Such a leak should have been possible using an etched track in mica, but none had been etched long enough to produce a suitably large (1-μm) hole that is of gigantic size relative to those shown in Figure 1-2, which are 100 times smaller. (A micron is 10^{-6} m, about 1% of the thickness of this page.). So Price and Walker (1962c) etched mica longer, and the results added two new dimensions – an enlargement technique and a scientific opportunity. After the track itself is rapidly etched away in mica, the walls of the etched holes continue to dissolve, but now at a much slower rate. The remarkable (yet, in retrospect, obvious) result is that tracks can be expanded in diameter to sizes that are readily visible in optical microscopes. Admittedly this result seems straightforward, but most new things are obvious once you think about them properly. The "obvious" opportunity took months to be realized!

The advantage in magnification is profound. Instead of the original 10-nm features such as those seen in Figures 1-1 and 1-2, tracks can be enlarged by etching to hundreds of microns. A few tracks per cm^2 can be seen instead of the $\sim 10^{10}$ cm^{-2} of, for example, Figure 1-1. Further, in Chapter 5, I will describe how we used a technique devised for another purpose to detect scarce tracks at the level of ~1 per m^2. In short, track densities can be measured over 14 orders of magnitude.

The other result of etching micas for an unusually long time was the finding of a practical way to observe built-in tracks in natural samples! Remarkably, these tracks have major scientific and economic uses in earth sciences and petroleum exploration. They are the key to the story to be told in Chapter 4.

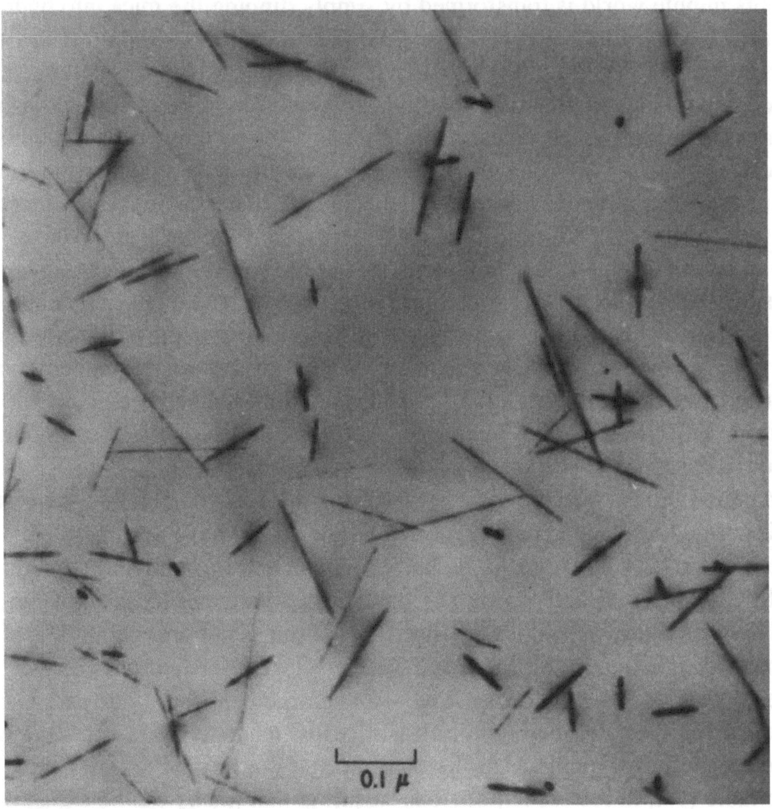

Figure 1-1. Nuclear tracks from the fission of uranium in mica seen in a transmission electron microscope by electrons that scatter off the strains around the tracks. The varying lengths are solely due to the many angles at which tracks cross the thin flake (Price and Walker 1962b).

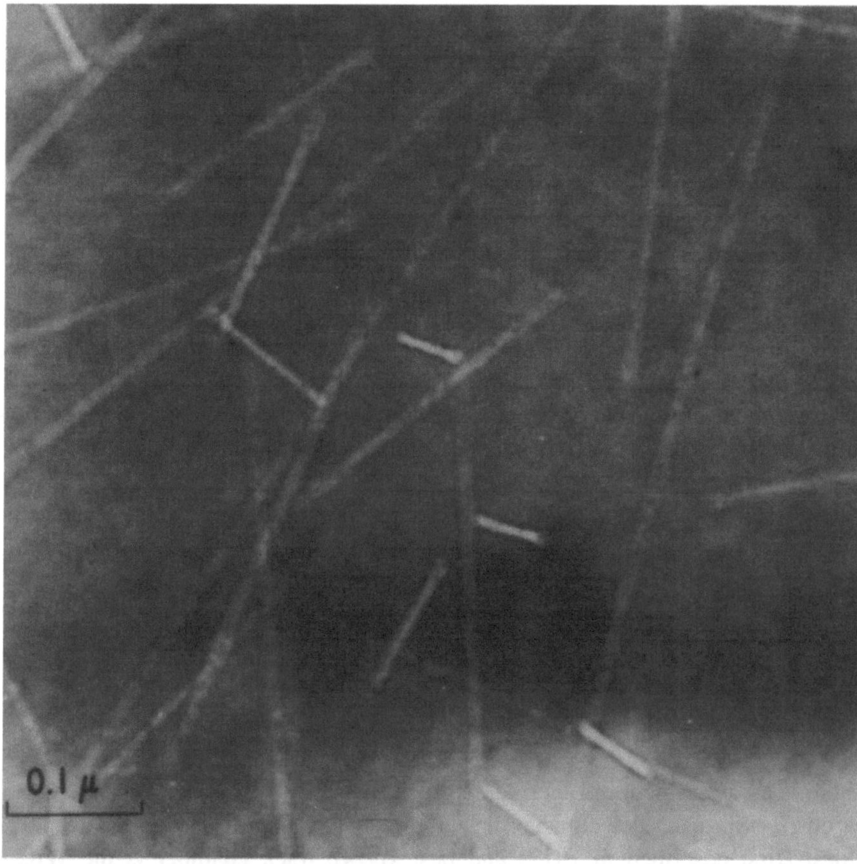

Figure 1-2. Tracks transformed into etched tunnels by immersion of mica in hydrofluoric acid for a few seconds. Thickness contrast shows 10 nm holes across the sheet (Price and Walker, 1962b).

Next come some of the basics of tracks—what solids record them, how they come about, and what controls their shapes and numbers. These factors will affect how they are used, as will be related in Chapters 2 through 6, which describe how controlled holes are drilled and used, how radon measurements help protect human health and give clues to processes within the earth, how ancient tracks test where oil can be found, what messages cosmic rays deliver, and many other unpredicted applications.

NUCLEAR PARTICLES

What are the particles that are of interest here? They are moving nuclei of atoms. Species can range from the lightest, hydrogen and helium, all the way through the chart of the elements to the heavy elements, such as thorium and uranium. The particles can be put in motion by manmade accelerators, by natural accelerators (processes at stars and between them), or by natural or induced nuclear disintegrations.

The best known of the natural decays produce alpha, beta, and gamma rays. Gamma rays are of the same family as X-rays, but are more energetic. They do not produce tracks in solids but can cause nuclear reactions that do. Beta particles are moving electrons. They too do not directly leave tracks in solids, but they can be seen in other, more complicated detectors. Alpha particles are merely fast-moving nuclei of helium. They do make tracks in many plastic solids. Radon detection is an especially important case. Because the radioactive gas radon decays by α emission, plastic detectors have given us a key tool for measuring radon.

A less well known type of decay is natural fission—the spontaneous splitting of a heavy atom such as uranium into two rather heavy parts, each roughly half of the mass of the original atom. In solids such *fission fragments* move 10–20 μm, and these particles are what are imaged so crisply in Figures 1-1 and 1-2.

OTHER NUCLEAR PARTICLE DETECTORS

There is an assortment of other detectors that display the paths taken by nuclear particles, with each detector having its special advantages and drawbacks. The oldest (first described in 1912) and best known track chamber is the Wilson cloud chamber. In it the pressure of the working vapor, for example, moist air, is lowered to the point of supersaturation, at which time ionization by moving particles can nucleate tiny droplets that display the paths of the particles. A bubble chamber is analogous: Here, within a liquid that is close to boiling, ionization produces lines of bubbles – displaying gas bubbles in a liquid, rather than the liquid droplets in a gas of the cloud chamber. Bubble chambers can be immense, one well-known example being the 15-ft (4.6-m) detector at the Fermi National Accelerator Laboratory near Chicago. Both types of devices have high sensitivity to lightly ionizing particles and give real-time, transient observations that need to be photographed to preserve their records. That characteristic may be compared to those of solid detectors – which accumulate particle tracks, for simpler display of infrequent events, but with little knowledge of when individual tracks were formed.

Spark chambers are somewhat similar. Ionization in the presence of appropriately high voltage nucleates electrical discharges between wires that are part of an array, and the paths can be recorded electronically as well as visually.

Most closely comparable to solid track detectors are photographic nuclear emulsions – photographic plates – the earliest imagers of nuclear radiation. Ionization sensitizes grains of AgBr that are the sensitive elements. Development of the film then reduces the sensitized AgBr to dark lines of silver particles. Like the solids, these plates are passive detectors, and they share in the attributes and difficulties of passive devices.

Each of these several detectors is capable of imaging less densely ionizing particles than can the solid track detectors, for example, electrons or mesons. But the solids have their own special virtues of detection efficiency, ruggedness, portability, and insensitivity to extreme conditions. They come in many varieties; hence, appropriate sensitivities can be chosen. They can be as tiny as an individual mineral grain or as large as a massive, many-square-meter array of plastic sheets. And as will be seen in Chapters 4 and 5, some natural solid detectors have existed since early in the history of the solar system. They can thus uniquely preserve records of ancient nuclear particle activity that preceded the existence of the other detectors.

WHAT TRACKS ARE SEEN IN PARTICULAR SOLIDS?

Track detectors as curiosities are transformed into useful tools once they are calibrated as to what nuclear particles create damage tracks and how efficiently they are displayed by chemical etching. The principles that I now discuss are simple, but the details (most of which I will omit) can be complicated.

What Fraction of Tracks Is Seen?

Of particle registration and etching efficiency, the latter is the easier to understand, being a purely geometrical effect. Figure 1-3 sketches how etching proceeds for two different tracks (in Figure 1-3A). The double dashed line for the left-hand track indicates that the extent of damage is more intense than for the right-hand, single-dashed track. That difference leads to more rapid dissolution (of velocity v_T) along the track than for the one on the right. The shapes of the etched tracks result from simultaneous etching of the undamaged material at a general rate v_G and the accelerated rate v_T along the track. The pits that result have V-shaped cross sections as sketched and therefore in three dimensions are conical.

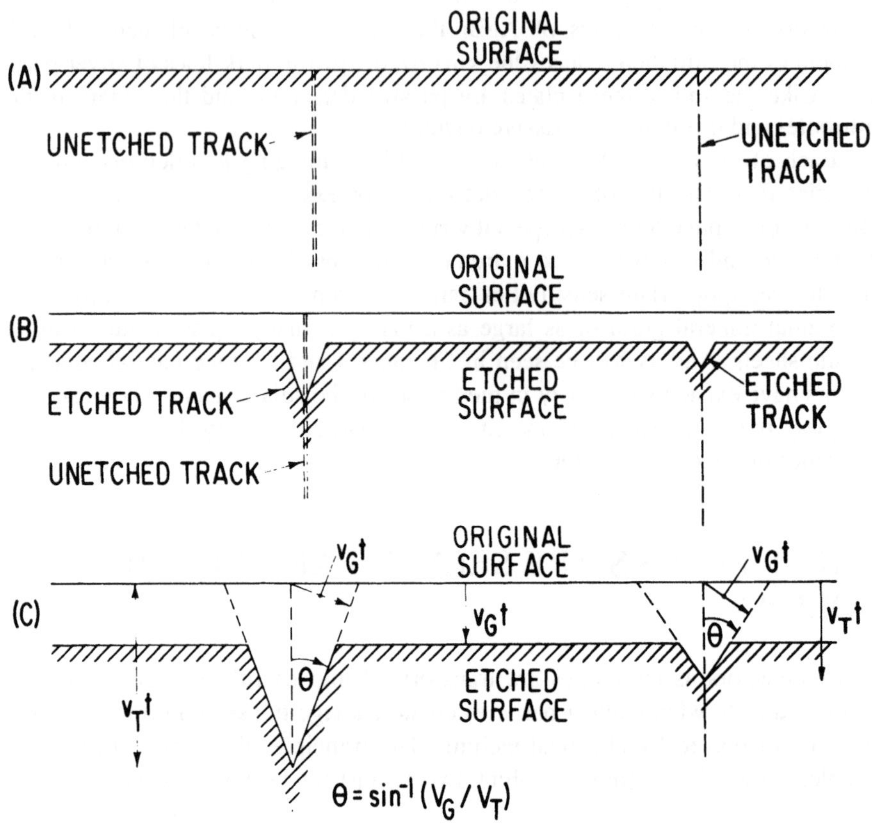

Figure 1-3. Track-etching geometry. Preferential etching along tracks at a rate v_T plus general etching at a lower rate v_G produce conical holes with a cone angle that can be shown to be $\sin^{-1}(v_G/v_T)$. The angle is larger and the pit is shallower for the less intense track on the right. A shows the unetched sample; B and C show the progressive effects of etching.

Figure 1-4 illustrates the typical case of an isotropic detector material with a clearly visible cone angle. Here tracks cross a thin sheet and are therefore etched from both sides as pairs of cones coming in point-to-point along individual tracks the two opposite surfaces, both of which were exposed to the hydrofluoric acid that was used as an etchant.

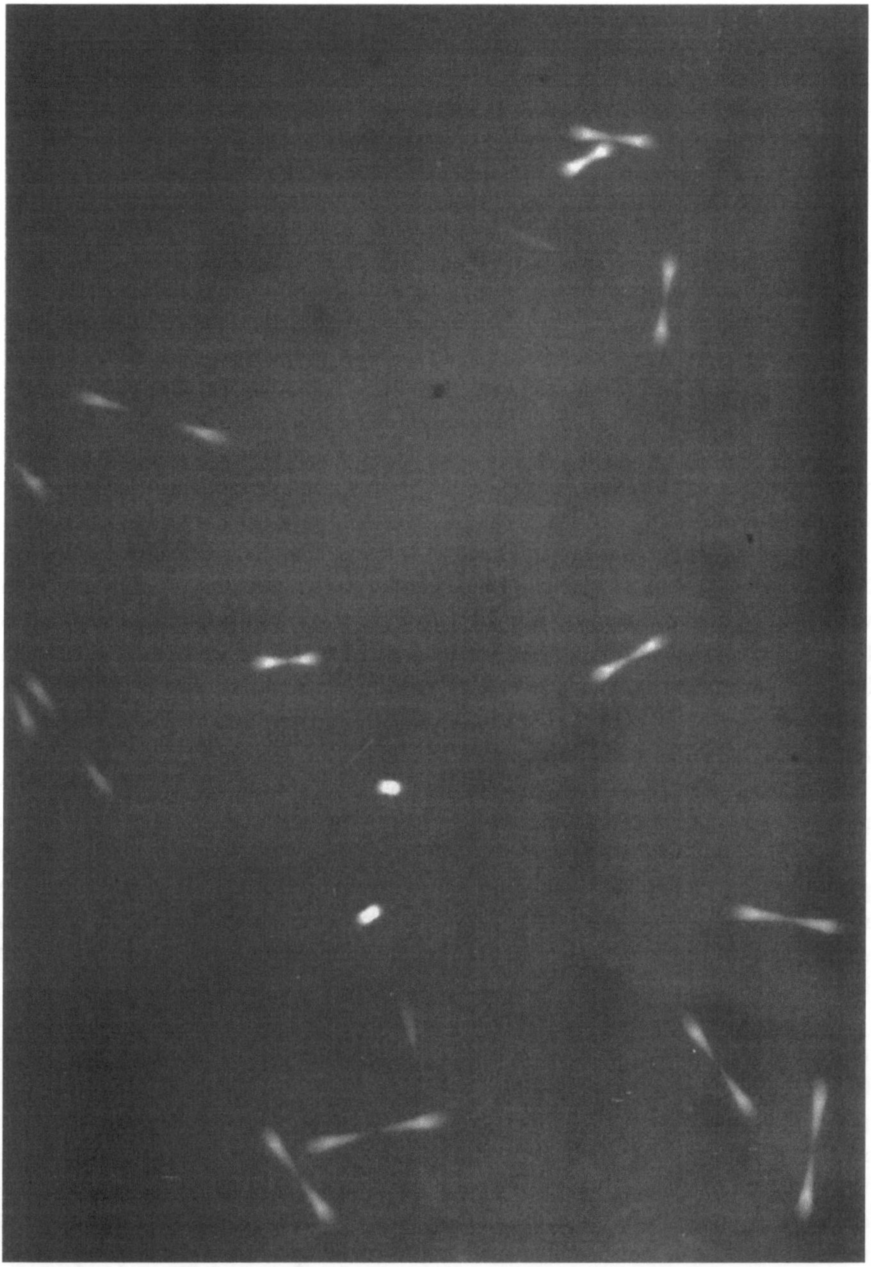

Figure 1-4. Tracks of finite cone angle through a thin sheet of phosphate glass. The bow-tie appearance comes from the separate etching of each track from the front and back of the sheet. The long dimension of the photo is 7.2 μm (Fleischer and Price, 1963).

Motor Boats and Tracks

A simple analogue that explains the V-shapes in Figures 1-3 and 1-4 is that of the bow waves of motor boats proceeding at constant velocity v_T through water in which the usual wave velocity is v_G. For larger v_T (i.e., a faster boat, as at left in Figure 1-3) the boat forges ahead, building a longer and therefore sharper V. In Chapter 5 we will see that when v_T varies along a track, interesting, highly useful deviations from the V-shape appear, deviations that make possible a major advance – identification of individual particles.

When θ (as defined in Figure 1-3) is zero, all tracks that intersect a surface are revealed, i.e., the etching efficiency is 100%. For finite θ some of the tracks from a randomly oriented distribution are lost if the etching along tracks lies entirely within a distance $v_G t$ of the surface. Within this slab of detector, the material where pits might have been is removed by general dissolution cutting into surface. A fraction that is given by $\sin\theta$ ($= v_G/v_T$) is lost, and therefore the efficiency is $1-\sin\theta$ [$= (v_T-v_G)/v_T$] (Fleischer and Price, 1964).

As we will see in Chapter 4, cone angles are also important in mineral dating and petroleum exploration because track shapes are changed in a recognizable way by heating.

Note that for tracks whose inclination to the surface is greater than θ, the efficiency is unity. Any external aligned beam of particles satisfying this condition will be fully displayed, *if the particles do form tracks*. In particular even if θ is large, all tracks at normal incidence to a surface will be etched. If a parallel beam of particles irradiates a sphere or a cylinder (Fleischer *et al.*, 1969) tracks of non-zero θ will appear on less than the illuminated half of the surface, the fraction being a direct way to measure that angle.

The particular micas that played such a vital role in the early discoveries are most exceptional materials in having essentially unit etching efficiency, zero cone angle, and immense anisotropy of etching rate. Etching normal to the layer planes is very nearly zero – leading to the observed etching efficiency. Etching in the plane is also very slow (though clearly measurable) and is much less than usual track etching rates, as evidenced by the uniform bores of the holes shown in Figure 1-2.

Etchants have now been established for a vast variety of glasses, plastics, minerals, and other crystalline solids (FPW75). It would be pleasing to be able to say that there is a direct way of identifying appropriate etches, but only broad principles exist. Reagents that slowly and uniformly dissolve away layers from a surface are appropriate. Adjusting temperature, concentration, and time so that a few microns of material are removed allows tracks to be seen in an optical microscope if present in sufficient numbers. For crystals, etches that are known to reveal other line defects, such as dislocations, are expected to do the job.

How Does Track Formation Occur?

Here we have a more profound question: Under what circumstances is an etchable track formed? To answer this question fully one must understand the mechanism (or more probably, mechanisms) of radiation damage formation and how the intensity of track damage varies with the nature of the particle and its speed. The problem has been attacked empirically, by amassing relevant experimental data and by physical modeling that allows interpretation of the data. Although mysteries remain (MRS95), the general picture is clear and widely but not universally accepted.

The Empirical Approach

The direct path to deciding what particles register is to irradiate samples of a solid with various particles at assorted energies and then use an etchant to reveal the tracks if they exist. This laborious procedure yields only a catalogue of decisions as to whether tracks are formed, each from an experimental result like that in Figure 1-5. If such data are all that is known, making predictions about other particles and energies is uncertain – informed guesswork at best. Adding intuition as to the physics that controls track formation gives a model that helps organize the data and allows trends to be recognized with fewer experiments.

And such experiments in the early 1960s were laborious. They meant traveling to an accelerator of heavy ions such as C, N, O, Ne, Ar... and being on call around the clock until a particular particle beam became available. It also included testing a beam to be sure it really had the ions that were injected and not merely contamination that happened to get through the acceleration system. We developed a first-look system using cellulose nitrate sheets (the most sensitive track detector at that time) and makeshift etching. Lacking a controlled etching bath at the accelerator, we dumped sodium hydroxide pellets into water in a beaker; the water promptly became hot by the heat of solution and was used immediately to etch the plastic for a few seconds – giving a quick but ill-characterized image of the beam size and intensity. Deferred maintenance made the ups and downs of beams unpredictable – in contrast to the orderly scheduling that can be done with better-funded, modern-day accelerators.

Our first hypothesis (guess) to model registration properties was that the threshold for track formation was specified by a particular level of energy deposition per unit distance (dE/dx). A charged particle loses energy by its positive charge exciting and ejecting the negatively charged electrons from atoms along its path as sketched in Figure 1-6). Our early experiments (Fleischer *et al.*, 1964) were scanty enough to be consistent with a sharp dE/dx level dividing zero from unit registration. But further experiments showed that a dE/dx threshold was fuzzy – it changed with the velocity of the particle and hence was not a true threshold. The left half of Figure 1-7 shows as an example that alpha particles of fixed dE/dx can either form a track or not, depending on velocity, and that the threshold (between open circles and filled circles) for heavier and faster ions is at a far greater dE/dx value than is reached by alpha particles. Hence a critical dE/dx must be rejected as defining when tracks will be produced.

Our second proposal, however, fits the results – as the right side of Figure 1-7 shows. The quantity involved – primary ionization – is the number of electrons per unit path length that are directly ejected by a particle that passes through the solid. This is a quantity that can be directly measured for gases, but usually is calculated for solids. Although it may appear that ionization would vary similarly to energy loss, the two are quantitatively distinct (as Figure 1-7 shows). Why ionization is appropriate will emerge shortly when track formation models are described. But thorough tests of the sort illustrated in Figure 1-7 have been done for three substances – muscovite mica, Lexan polycarbonate, and cellulose nitrate. Each gave similar results – but with distinct values of the ionization threshold, i.e., different sensitivities.

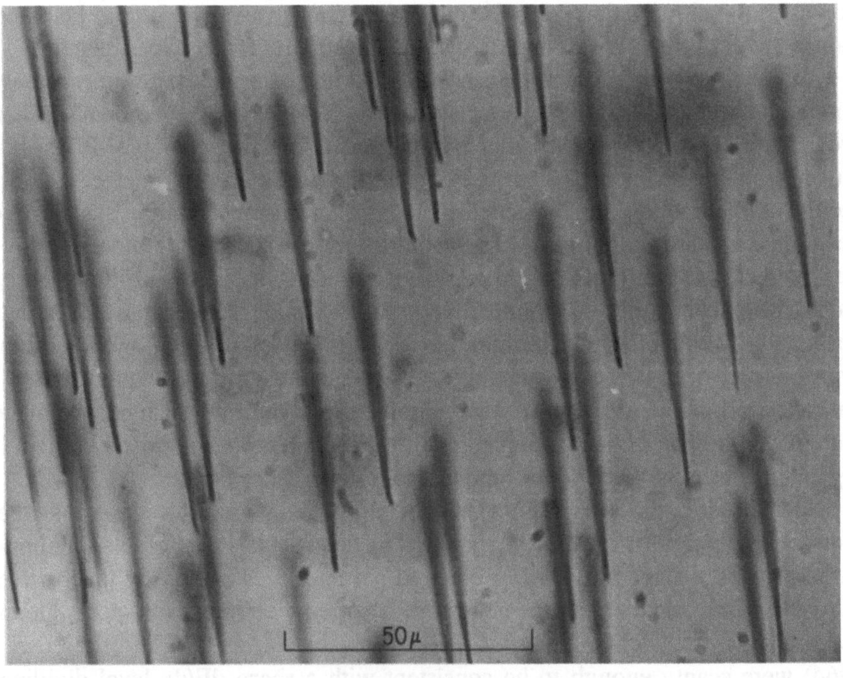

Figure 1-5. A test of whether 139 MeV ^{32}S ions register in Lexan polycarbonate plastic. Here the sulphur ions were accelerated to one-tenth the velocity of light. Clearly the beam does register here, and each particle leaves a track. Transmitted light photograph, focused on the ends of the tracks inside the plastic (Fleischer *et al.*, 1964).

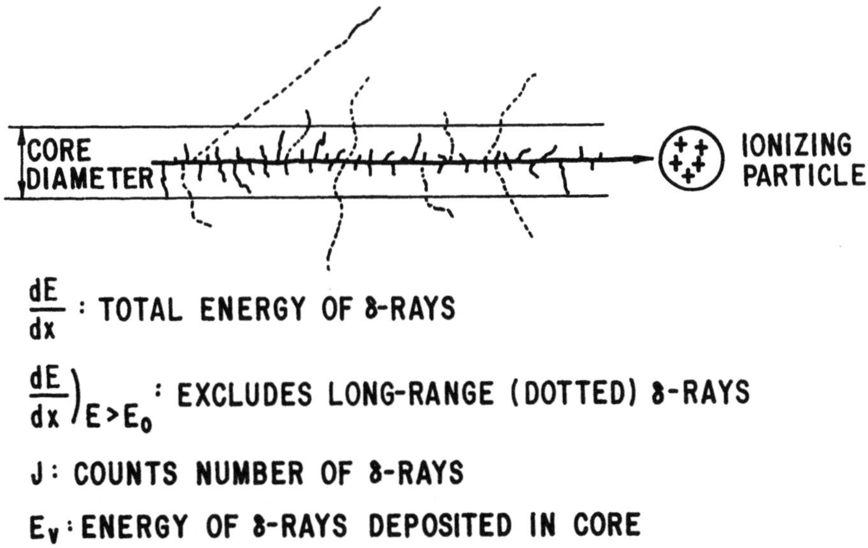

Figure 1-6. Sketch of paths of ejected electrons along the path of a heavy ion.

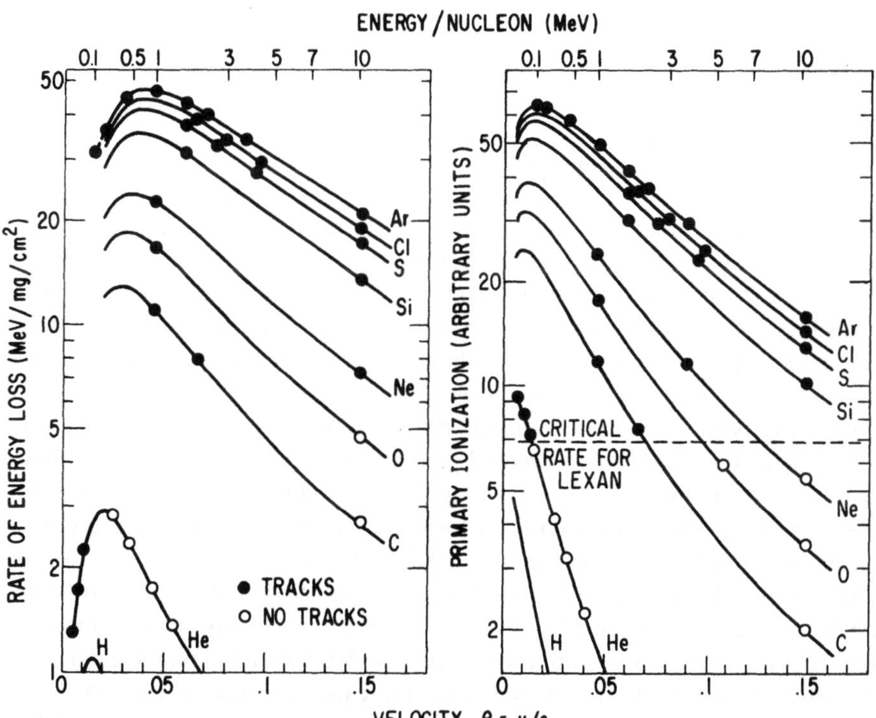

Figure 1-7. Possible criteria for track formation compared to experimental results for the plastic Lexan. A single threshold value using energy deposition rate is not possible (left), but for primary ionization rate (right) a single, well-defined threshold does exist (the dashed line). Open circles show where no tracks were found, filled circles where all particles formed tracks (Fleischer *et al.*, 1967).

Let's consider why energy loss and ionization vary as they do with velocity and ion species. Except at very low velocities, energy loss of ions is primarily by ionization, transferring energy by exciting, or ejecting, electrons close to the projectile's path. Coulomb attractions between electrons and the moving, positively charged particle yank electrons loose and propel them across and away from the ion path, and they in turn eject other electrons and lose energy - mostly very close to where the ion passed (see Figure 1-6). The electrons that normally are attached to a moving atom will be stripped off by collisions as it penetrates a solid, as the atom velocity exceeds orbital velocities of the bound electrons, and it thus becomes an ion.

The shapes of the curves in Figure 1-7 are determined by the interplay of two factors, the charge on the ion and the impulse it gives electrons. The impulse is proportional to the time the ion is close to an electron, which is inverse with velocity. It is this factor that gives the general (at first thought) counterintuitive decrease in energy loss and ionization with increasing velocity. But as an ion slows, it progressively re-acquires electrons to replace those that had been stripped off. This process lowers its net charge, and hence both ionization-related energy loss and primary ionization decrease toward very low energy. Each quantity goes to zero once the ion becomes electrically neutral. To be neutralized, an ion must be very slow, and such atoms are not known to produce tracks.

HOW TRACKS ARE FORMED

Although not essential to understanding how to use tracks, as described in the following chapters, it is interesting to consider how tracks come about. Those who are most interested in uses of tracks may skip to page 21, Track Stability. A plausible mechanism for track production must both fit the results in Figure 1-7 and agree with a wide spectrum of other experimental results. One of the most striking sets of results (summarized by Table 1-1) is the list of solids in which tracks have been seen or not seen after diligent effort. Basically the result is that good conductors usually don't record particle tracks, but insulators do. Another critical observation comes from heating solids that contain tracks and observing the time and temperature conditions for repairing the radiation damage ("track annealing"). This process is thermally activated and is characterized by activation energies (FPW75) of a few electron volts, values that are 1) typical for the random (*diffusive*) motion of atoms in solids and 2) much greater than those for moving electrons around within solids. Although reasonably complete data such as those presented for Lexan in Figure 1-7 exist for only a few substances, threshold values are known for dozens of solid track detectors (Fleischer, 1981). Explanation of the sequence of sensitivities of different substances would be an important element of an acceptable model of track formation.

Table 1-1. Relation of Track Formation to Electrical Resistivity

Solids	Resitivity Range (ohm-cm)
I. Track-Forming	
Insulators	
Silicate minerals	10^{6}–10^{20}
Alkali halides	
Insulating glasses	
Polymers (plastics)	
Poor Insulator	
MoS_2	3,000–25,000
Semiconductor	
V_2O_5 glass	2,000–20,000
II. Non Track-Forming	
Semiconductors	
Germanium	10–2,000
Silicon	
Metals	
Aluminum	10^{-6}–10^{-4}
Copper	
Gold	
Platinum	
Tungsten	
Zinc	

The challenge, then, is to find a mechanism in which the threshold for producing tracks depends on loosening electrons, but which causes damage that consists of atomic defects. And a full understanding of the mechanism should also allow sensitivities of different substances to be ranked.

The process that we named the "ion explosion spike" (Fleischer *et al.*, 1965) is sketched in the upper half of Figure 1-8. Imagine that a multiply charged ion has moved through a perfectly ordered crystal from right to left, ejecting a good many electrons from near its path. The ions left behind, being of like sign, repel each other strongly, propelling one another away from the original trajectory, in a miniature explosion that creates many displaced atoms and atoms in interstitial positions – leaving behind a region with a high concentration of vacant lattice sites. Because most of the solids of interest are ordered arrays of more than one type of atom, interchanges of atoms of unlike size and chemical properties further raise the catastrophic nature of the lattice disruption from what is sketched.

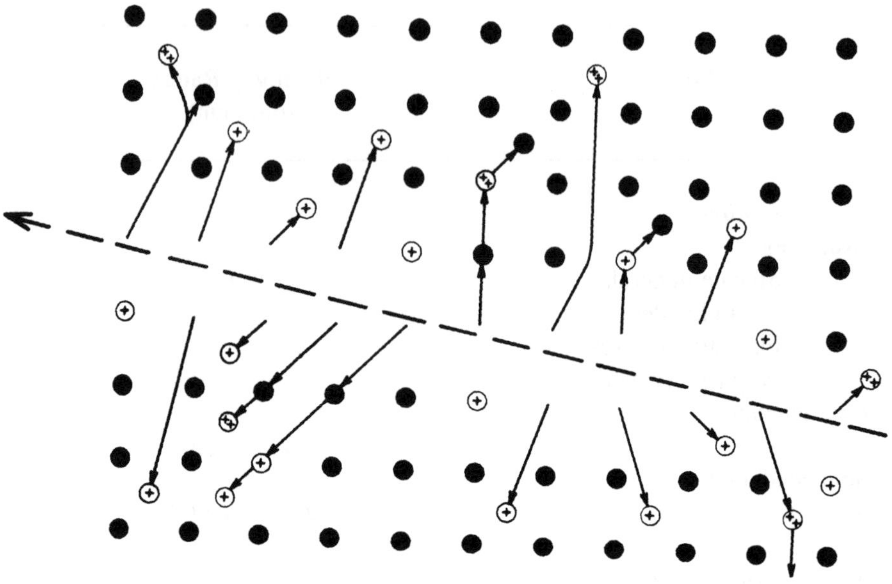

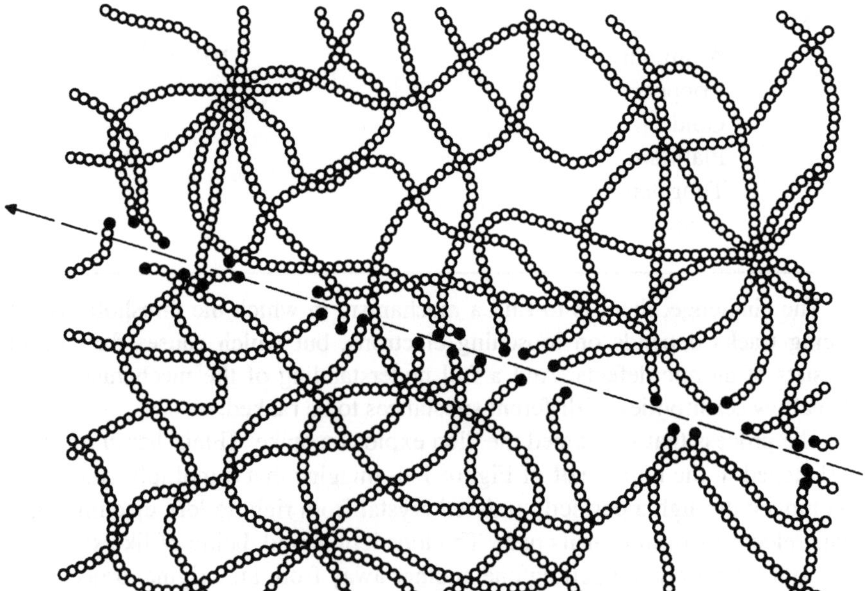

Figure 1-8. Track formation: a) The ion explosion spike mechanism. Electrons are ejected by a moving charged particle. The remaining positively charged region is unstable and the coulomb (repulsive, electrostatic) forces lead to atomic disorder. b) In plastics, chain breaks allow preferential etching at a lower damage intensity.

Note that this physical description fits two of the major requirements: the damage 1) occurs as a result of *ionization* closely along the particle trajectory (primary ionization) and 2) produces atoms in wrong locations, so that *atomic* diffusion is required to repair the damage by allowing the atoms to return to sites where they fit best. That the process of track annealing will turn out to be complex when looked at carefully is hardly surprising since the motion of more than one atomic species of displaced atom is involved, and initially the disturbed atoms can be part of a variety of disorderly atomic groupings. Once the gradual healing of tracks by heat has been calibrated in one of several ways to be described in Chapter 4, this complexity is useful both scientifically and technologically to learn how temperatures have varied over time in the geological past. Note also that disordering normally produces a volume expansion since atoms are not as tightly bonded to one-another as in the stable, original structure. The resultant volume expansion within the solid produces the lattice strains that imaged the tracks in mica in Figure 1-1.

The idea that multiple ionizations can lead to electrostatic ion repulsions is a massive scaling-up of an ionization-produced atomic process that was described by Varley (1954): He suggested that stripping an electron off of a single negative ion in an ionic crystal creates an unstable, positively charged region. Young (1958) adapted this idea to explain his tracks in LiF. His is qualitatively the same process that Price, Walker, and I later elaborated and made semiquantitative to explain and systematize the track-etching behavior of many materials.

Knorr (1964) provided a remarkably graphic view of the effects of the coulomb explosion. Figure 1-9 is his photo of the surface of a LiF crystal in which two heavy ions had approached and entered the surface at glancing angles. The particles (fragments from nuclear fission of uranium) moved from left to right—H to J′ and F to G′. From H to I the ionized region was at the surface so that the upward ejection was not confined by material above, and the escaping debris excavated a 140-Å-wide groove. From I to the step (an irregular white band) the track material was confined, but just barely. The bulge from the volume expansion of disordering is visible. The particle emerged from near the bottom of the step close to the level of the surface on the right of the step and plowed another groove to position I″ and created a second bulge from I′ to J′. The reader can compose a similar, but slightly different scenario for the second particle.

The idea of extreme disorder at the core of the track, such as the upper portion of Figure 1-8 implies, was what Price, Walker, and I had originally imagined – a glass-like, i.e., noncrystalline, region making up the rapidly etching part of the track. This idea briefly impeded our early progress at showing tracks can exist in diverse materials. You don't expect tracks in glass if the stuff of tracks is itself glass. Happily, from Berkeley Price smuggled into the lab a weak californium 252 radioactive source, which gives off fission particles. Its convenience made testing for tracks less time-consuming than debating what tracks were atomically. So a what-the-heck-let's-try-it experiment showed that glass recorded tracks, and another showed the same for another class of disordered materials – plastics. Walker, who at the time was on leave working in France on other aspects of tracks, was incredulous when I wrote him about these new detector materials. But the experiments were too easy and direct for him to remain skeptical for long. The message we finally unraveled about glasses is that, although the tracks in crystals and glasses may indeed be glass-like, there are many varieties of glasses, and the most unstable are likely to etch the most rapidly.

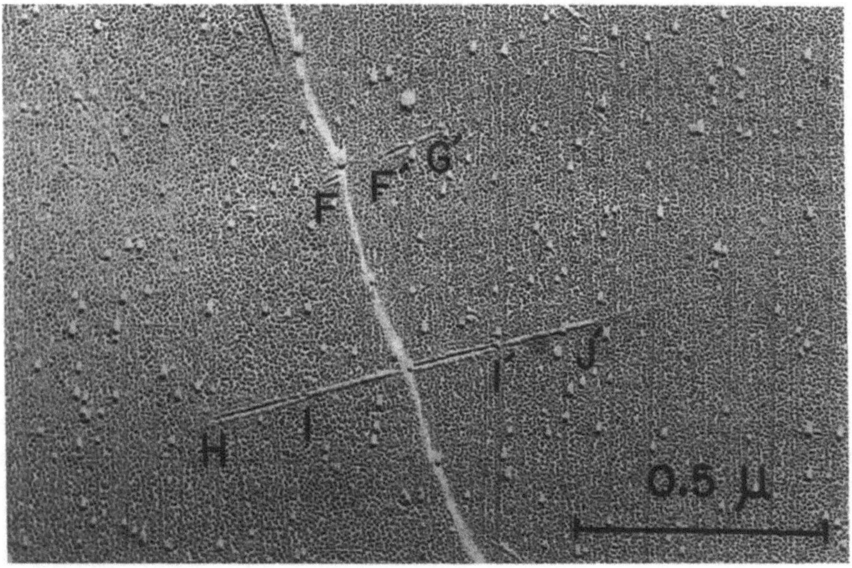

Figure 1-9. Nuclear track groove and bulge in a lithium fluoride crystal with a step. A fission particle at near-glancing-angle incidence caused the ion repulsions that created a groove HI (by ejecting ions from the surface) and then a bulge where the disordered region is contained within the solid. Coming out at the edge of a step, the particle approached the next surface and repeated the process (Knorr, 1964).

Detector Thresholds

Some understanding of thresholds can be achieved by thinking a bit further about Figure 1-8 (top). The damage occurs when adjacent atoms are converted into ions of positive charge. But if ionizations by the moving particle are too infrequent, the ions will be spaced so far apart that the coulomb (electrostatic) forces are too slight to displace ions from their lattice sites. Calculations of the ionization rate for ions that are at threshold for muscovite mica show it to be about one ion per atomic layer – what one would expect intuitively as a minimum to produce a continuous trail of damage. Remember, for muscovite mica the damage must be continuous, since the layer planes are not dissolved unless there is some source of preferential chemical attack. For typical etching times, a discontinuous track in mica will not etch past the first gap in the damage.

In glasses the etchant that dissolves the tracks also dissolves undamaged material, although less rapidly. Thus a continuous track is not a requirement, since slower dissolution will eventually bridge the gaps and reach the next sites of accelerated etching. As a result, most of the glasses are more sensitive particle detectors than the micas.

Most minerals, however, are less sensitive than the micas. An extension of our model relates the difficulty of producing damage to the "stress ratio," which is the ratio of the bond strength that holds the solid together (measured by the elastic moduli) to the coulombic interatomic repulsions (which tend to blow the solid apart; they rise inversely with a power of the interatomic spacing and with the shielding provided by the dielectric constant). As Table 1-2 shows, this ratio does broadly group detectors by their observed thresholds for particle detection, with those with the highest stress ratios being the least sensitive detectors.

Life is not totally simple, however. The plastic (polymeric) detectors are a special case. Being made of long-chain molecules, they are susceptible to effects of chain breaks, which can be created at considerably lower energy transfers to electrons (2–3 electron volts [eV]) than are needed in inorganic solids for the lowest energy ionization processes (10–15 eV). Chain breaks in turn lower the molecular weight and allow more rapid chemical attack at the increased number of chain ends (Fleischer *et al.*, 1965). Figure 1-8 (bottom) illustrates how a track will include a high concentration of molecular terminations, and justifies further why polymers are the most sensitive of the track detectors. There are great gaps in our understanding of polymers. For instance, it is unclear why some (for example, a natural plastic, amber) are among the least sensitive detectors. O'Sullivan *et al.* (1982) showed that particle detection sensitivity correlates with the ease of producing chain breaks for a number of polymers. Other factors that may be important are additives (of which there are many), fine-scale uniformity, ability to crystallize partially, and relative stability.

The ion spike model tells us why good electrical conductors normally do not allow tracks to form. A brief but finite time is needed for the electrostatic repulsion of adjacent positive ions to repel one another into the lattice. If the ions are

Table 1-2. Relation of Registration Thresholds to Calculated Stress Ratio

Detector class	Detector groups of similar registration thresholds (Highest threshold at top)	Average stress ratio
Crystals	Hypersthene Olivine	4.5
	Labradorite Zircon	1.4
	Diopside Augite	1.3
	Oligoclase Bytownite Orthoclase Quartz Micas	0.5
Inorganic glasses	Silica Flint Tektite	0.7
	Soda lime Most phosphate	0.5
Plastics	Group of 14 plastics	0.01

neutralized by re-acquiring electrons in a shorter time than is required for the ions to vault into the lattice, the ions cease to exist, their electrostatic forces are no more, and the potential track is quashed.

Summaries of recent work (MRS95) showed surprising and important results that appear to violate the simple division of substances by electrical resistivity like that in Table 1-1. Tracks have been seen in oxide superconductors, intermetallic compounds, and in at least one metal (titanium). In the oxide superconductor $Bi_2Sr_2CaCu_2O_x$ ("Bi-2212") tracks of 300-MeV lead ions were clearly displayed by Provost *et al.* (1995) using transmission microscopy, and Barbu *et al.* (1995) showed tracks of particles of comparable ionizing power in the intermetallics $NiZr_2$ and NiTi and the metal Ti (the Ti irradiated at 20K). Resistivities of such materials lie in the range 40 to 2,000 μ ohm-cm, and hence overlap the non-track-forming semiconductors in Table 1-1.

It is premature to conclude that a major problem exists in deciding which materials allow tracks to form. Table 1-1 was constructed using results from fission particle irradiations, which as yet have not produced tracks in these new track detectors. The ionizing power of the ions that formed tracks is 2 to 3.5 times that available from fission fragments, so that use of these particles – now available in accelerators – might shift where the transition from no tracks to tracks occurs for some of the near-threshold material in the table. For the superconducting materials it has been shown that both in the oxides such as Bi-2212 (Luborsky *et al.*, 1991) and certain intermetallics (Bean *et al.*, 1966) fission fragments produce damage that markedly enhances superconducting properties, particularly at high-fields, even though tracks could not be observed. Those results suggest that fission fragments were close to but did not quite reach the threshold for producing continuous, etchable tracks.

TiNi, Ti, and $NiZr_2$ are special materials in having solid-state phase transitions as stress or temperature increases – that is, a change from one crystal structure to another or to a disordered solid. Their presence implies that those materials are close to instability and therefore might be expected to be more sensitive to particle-induced disturbances. $NiZr_2$ can be made amorphous by mechanical deformation; NiTi, a shape-memory alloy, undergoes a martensitic transformation (a phase change by massive shear) at 35°C; and Ti has an allotropic transformation from a hexagonal structure to a cubic one at 882°C. Barbu *et al.* (1995) raise two possibilities as to special features of track formation, one involving soft phonon modes, the other shock waves. In support of the shock wave possibility we noted (Fleischer *et al.*, 1965) that, in addition to the primary ionization loss to electrons, an ionizing particle also transfers energy to the ion cores, but in far lesser amounts. For the ions now being used (for example, 1 GeV U) a titanium atom that is passed by the projectile at a distance of 0.8 Å is given 0.3 eV, insufficient to displace a single atom but nevertheless 12 times the thermal energy. Hence there will be a cylindrical elastic pulse from the passage of such an ion, perhaps enough to nucleate a phase transformation in cases where such a transformation requires not too great a stress level.

TRACK STABILITY

The great durability of tracks in solids is noteworthy, since it allows them to persist under adverse conditions of temperature, pressure, and such, that would degrade the workings of other nuclear detectors. This stability has permitted tracks to be identified that formed near the beginning of the solar system in meteoritic minerals, and dosimetry measurements to be made in an intense background of less concentrated ionizing radiation. In one case a detector package that was lifted to record cosmic rays near the top of the atmosphere was accidentally released and free fell from 100,000 feet into a swamp. The particle tracks survived unaltered, although bugs were found in the system (literally).

A wide range of environmental effects have been assessed in track detectors, including temperature, pressure, shock, humidity, and ionizing radiation. Specifics can be found in our book (FPW75), but the broad features are these: Many minerals store tracks tenaciously, but others only weakly. Hydrostatic pressure to 80 kbar has only a slight effect, but shock deformation using comparable pressure can erase or obscure tracks in inorganic detectors. As an example, Figure 1-10 shows how tracks are lost close to an underground nuclear explosion, but preserved (mostly) at >30 meters distance. The 50% loss point, as drawn, corresponds to a shock pressure of roughly 50 kbar.

Tracks in minerals and glasses ignore vast doses of electrons and of ultraviolet (UV) radiation and show no effects of external oxygen exposure, but plastics are another story entirely. Effects of UV exposure, humidity, oxygen environment, and temperature abound. Careful work requires knowledge of these possible factors and control of at least some of these parameters, depending on the particular plastic.

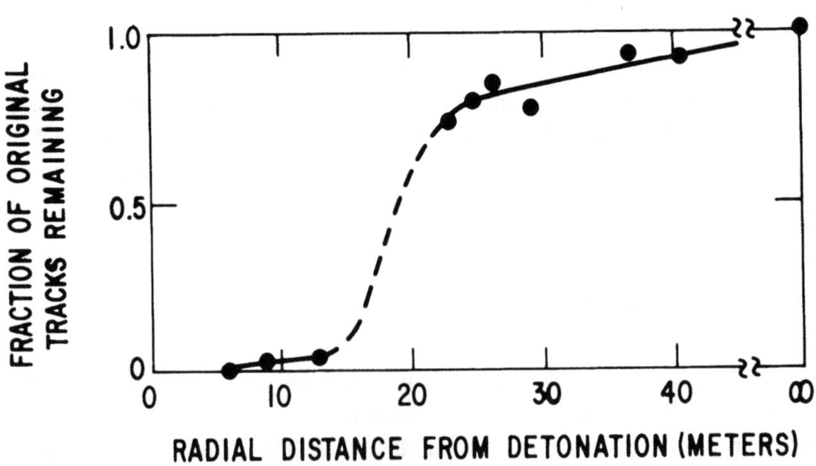

Figure 1-10. Track erasure in the mineral apatite from rocks that were shocked at the Hardhat underground nuclear test (Fleischer *et al.*, 1974).

CONCLUDING COMMENTS

The reader will recognize that a great deal is known about the properties of a wide spectrum of track detecting solids, and a broadly applicable semiquantitative model fits a great many of the diverse properties. Significant gaps in the understanding of the detailed mechanism remain, providing workers in this field with ample opportunities for further discoveries. The gaps in understanding, although scientifically significant, have nevertheless not seriously hampered using tracks in a multitude of areas of science and technology, as Chapters 2 through 6 will show.

The importance of the science described here is that ordinary, everyday solid materials have abruptly been transformed into detectors of nuclear particles. A piece of glass or a mineral may be put into an accelerator beam and nuclear reactions recorded, or natural solids may be developed to reveal historic nuclear events and particles from cosmic accelerators. And in the next chapter we will see that the results enticed Price, Walker, and me on a "holey" quest.

FURTHER READINGS ON THE NATURE OF TRACKS AND OF TRACK ETCHING

A general reference:

DB87 Chapters 1 through 5 of *Solid State Nuclear Track Detection*, S. A. Durrani and R. K. Bull, Pergamon Press, Oxford, 1987, pages 1–113.

A popular-level survey:

FPW69 "Nuclear Tracks in Solids," R. L. Fleischer, P. B. Price, and R. M. Walker, *Scientific American* **220**, 30-39 (1969) June.

For broad general coverage:

FPW75 Chapters 1 and 2 of *Nuclear Tracks in Solids*, R. L. Fleischer, P. B. Price, and R. M. Walker, Univ. of California Press, Berkeley, 1975, pages 3 to 118.

On the many materials in which tracks form:

MRS95 *Ion Tracks in Solids* issue of *Materials Research Society Bulletin*, Dec. 1995, **20** (12), 17–41.

For track fundamentals, track etching:

S90 Part I of *Ion Tracks and Microtechnology*, R. Spohr, Vieweg, Braunschweig, 1990, pages 14 to 182.

REFERENCES

A. Barbu, H. Dammak, A. Dunlop, and D. Lesueur (1995), "Ion Tracks in Metals and Intermetallic Compounds, " *MRS Bull.* **20** (12), 29–34.

C. P. Bean, R. L. Fleischer, P. S. Swartz, and H. R. Hart, Jr. (1966), "Effect of Thermal-Neutron Irradiation on the Superconducting Properties of Nb_3Al and V_3Si Doped with Fissionable Impurities," *J. Appl. Phys.* **37**, 2218–1114.

R. L. Fleischer (1981), "Nuclear Track Production in Solids," in J. W. Christian, P. Haasen, and T. B. Massalski, eds., *Progress in Materials Science*, Chalmers Anniv. Vol., pp 97–123.

R. L. Fleischer and P. B. Price (1963), "Charged Particle Tracks in Glass," *J. Appl. Phys.* **34**, 2903–2904.

R. L. Fleischer and P. B. Price (1964), "Glass Dating by Fission Fragment Tracks," *J. Geophys. Res.* **69**, 331–339.

R. L. Fleischer, P. B. Price, R. M. Walker, and E. L. Hubbard (1964), "Track Registration in Various Solid State Nuclear Track Detectors," *Phys. Rev.* **133A**, 1443–1449.

R. L. Fleischer, P. B. Price, and R. M. Walker (1965), "The Ion Explosion Spike Mechanism for Formation of Charged Particle Tracks in Solids," *J. Appl. Phys.* **36**, 3645–3652.

R. L. Fleischer, P. B. Price, R. M. Walker, and E. L. Hubbard (1967), "Criterion for Registration in Dielectric Track Detectors," *Phys. Rev.* **156**, 353–355.

R. L. Fleischer, P. B. Price, and R. T. Woods (1969), "Nuclear Particle Track Identification in Inorganic Solids," *Phys. Rev.* **88**, 563–567.

R. L. Fleischer, R. T. Woods, H. R. Hart, Jr., P. B. Price, and N. M. Short (1974), "Effect of Shock on Fission Track Dating of Apatite and Sphene Crystals from the Hardhat and Sedan Underground Nuclear Explosions," *J. Geophys. Res.* **79**, 339–342.

T. G. Knorr (1964), "Fission-fragment Tracks and Directional Effects in the Surface of LiF Crystals," *J. Appl. Phys.* **35**, 2753–2760.

F. E. Luborsky, R. H. Arendt, R. L. Fleischer, H. R. Hart, Jr., K. W. Lay, J. E. Tkaczyk, and D. Orsini (1991), "Critical Currents After Thermal Neutron Irradiation of Uranium Doped Superconductors," *J. Mater. Res.* **6**, 28–35.

D. O'Sullivan, P. B. Price, K. Kinoshita, and C. G. Wilson (1982), "Predicting Radiation Sensitivity of Polymers," *J. Electrochem. Soc.* **129**, 811–813.

P. B. Price and R. M. Walker (1962a), "Observation of Charged Particle Tracks in Solids," *J. Appl. Phys.* **33**, 3400–3406.

P. B. Price and R. M. Walker (1962b), "Chemical Etching of Charged Particle Tracks," *J. Appl. Phys.* **33**, 3407–3412.

P. B. Price and R. M. Walker (1962c), "Observation of Fossil Particle Tracks in Natural Micas," *Nature* **196**, 732–734.

J. Provost, Ch. Simon, M. Hervieu, D. Groult, V. Hardy, F. Studer, and M. Toulemonde (1995), "Swift, Heavy Ions in Insulating and Conducting Oxides: Tracks and Physical Properties," *MRS Bull.* **20** (12), 22–28.

E. C. H. Silk and R. S. Barnes (1959), "Examination of Fission Fragment Tracks with an Electron Microscope," *Phil. Mag.* **4**, 970–971.

J. H. O. Varley (1954), "A Mechanism for the Displacement of Ions in an Ionic Lattice," *Nature* **174**, 886–887.

D. A. Young (1958), "Etching of Radiation Damage in Lithium Fluoride," *Nature* **182**, 375–377.

Chapter 2
HOLE ENGINEERING—FROM FILTERS TO COUNTERS

...to assume a pleasing shape
W. Shakespeare, in *Hamlet*, Act II, Scene 2

Like a vacuum, a hole is not much. But holes of controlled geometry have many uses, both scientific and technological. And commercial applications abound.

Holes made using particle tracks were the first practical application of tracks that arose at General Electric, the purpose being to create a new type of filter with holes of great regularity and widely adjustable sizes. The principle is straightforward. A beam of track-forming particles is directed across a thin sheet of a track-detecting material. Etching then drills holes through the sheet – holes of nearly identical size and shape, their number dictated by the dose of particles and their size specified by the etching time.

The first such filter was made laboriously from mica (Fleischer *et al.*, 1963), with results such as those shown in Figure 2-1. The directional character of the chemical dissolution of mica is made obvious by the diamond-shaped holes. Our first process of irradiation was awkward. We used fission of ^{235}U to produce the particles, a random process that spews fission fragments in all directions. Lining up the tracks so that they could cross the mica sheet therefore required separating the uranium fission source from the mica target and evacuating the intervening space. And since the fission was to be induced by neutrons, the assembly was enclosed in a capsule small enough to go into a nuclear reactor. So we had an aluminum container, with a thin film of uranium across one end and a sheet of mica at the other. It was welded shut and then pumped out through a small aluminum pipe that was then pinched off to provide the final seal against air leaking in. We hoped. Each exposure in the reactor was followed by a period of suspense while we tested whether the tracks were where we wanted them, or whether we had to try again.

Irradiations to produce filters soon became far easier once we found that californium sources were available. Californium 252 undergoes spontaneous fission and can be obtained as a minute layer sputtered on the center of a platinum disk. Irradiations can be done in the laboratory, where the target material can be placed in a bell jar with the source, and the jar pumped out.

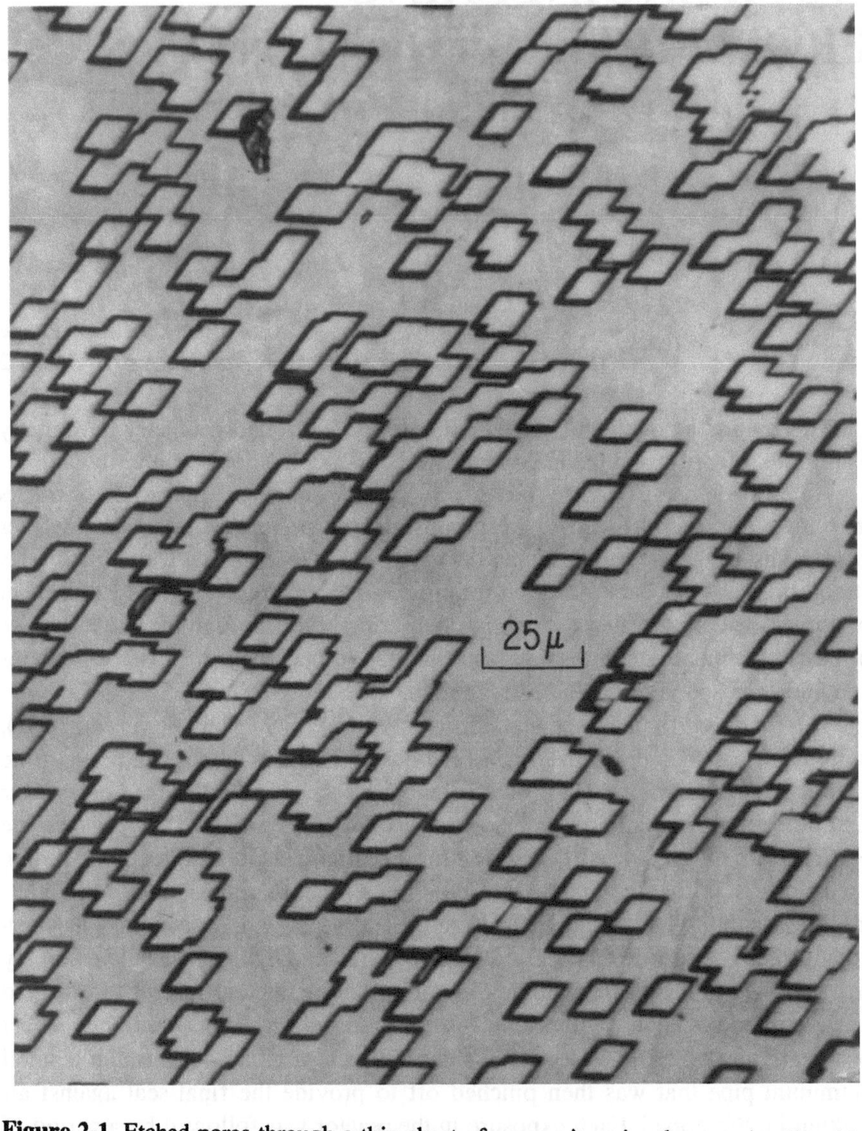

Figure 2-1. Etched pores through a thin sheet of muscovite mica that was irradiated by fission fragments from ^{235}U in a nuclear reator. The diamond shapes result from the dissolution anisotropy of the crystal structure (Fleischer *et al.*, 1963).

Originally mica was used for filters because it was the first solid detector we knew of, but it suffers from brittleness that made the delicate, 10-μm sheets difficult to handle. A more rugged and practical material is Lexan polycarbonate, which we soon learned was readily etched with caustic solutions (Fleischer and Price, 1963). Figure 2-2 shows holes in Lexan compared to a conventional cellulosic filter. It illustrates another feature of most polymeric substances, isotropic etching, which gives nicely circular holes.

The event that sparked us to continue to work on filters was initiated by a physician, Dr. S. M. Seal. From the report of Suits and Bueche (1967), which includes the quotation that appears before the preface on page v: "One day in November 1962, Price received a phone call from Dr. S. H. Seal of the Sloan-Kettering Memorial Cancer Center in New York City. Seal was attempting to isolate and detect cancer cells in blood by filtering the blood through sieves that would hold back the larger, more rigid cancer cells while allowing others to pass through. Filters he had tried previously were unsatisfactory, and he was searching for more suitable types. By word of mouth he had heard of the sieves that could be made with fission fragments. Price agreed to etch holes 3 to 4 microns in size in 1-inch-diameter pieces of mica. But when Seal tried them they broke, and he asked for more."

After we found that plastics could be used, a new ballgame could begin. "Plastics had several advantages; one of the most important was the fact that they could be handled and used more easily than large sheets of glass (which was also found to give etchable tracks) or mica. Early in May, filters were on their way to Seal, and his experiments were promising. The cancer-bearing blood flowed through the filters leaving the cancer cells deposited on the transparent plastic and allowing them to be strained *in situ* and examined under a microscope."

The article that included the etched filters in Figure 2-2 (Fleischer *et al.*, 1964) – with E. M. Symes – was the first open announcement of the possibility of making precisely sized filters suitable for cell separation (about 5 μm diameters), and we expected (and later received) much interest from the biological community, with which we were not familiar; and, therefore, we paid special attention to get things right. Because this was a new field to us, we asked for comments on the draft of our manuscript from Seal and a consulting physician who worked at our laboratory, H. M. Rozendaal. We received only a single comment from each: "[In your acknowledgments] refer to me as *Doctor* S. M. Seal" and "You have misspelled Rozendaal." Dr. Seal was later to compound the problem in a paper he wrote by thanking H. M. Romensdahl and E. M. Synes.

The follow-up to the potential filter business at the General Electric Research Laboratory was the creation by Dr. C. A. Bruch in the summer of 1963 of a pilot-plant operation to manufacture polycarbonate filters ("PIE filters," for "plastics, irradiated and etched"), and later, in 1964, of a full commercial operation at Vallicitos Nuclear Laboratory, of General Electric's Nuclear Energy

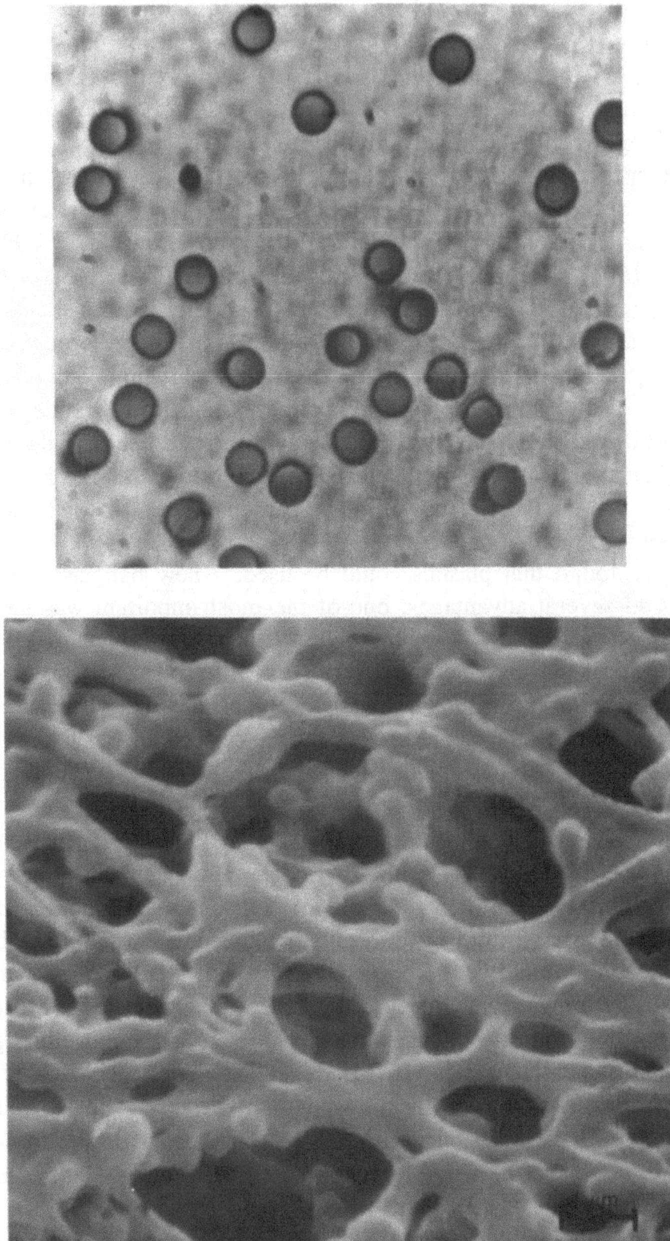

Figure 2-2. Comparison of filter types. Top: Etched uniformly circular 5-μm holes through a film of polycarbonate plastic that was irradiated with fission fragments from ^{252}Cf in an evacuated bell jar.. Bottom: Cellulose membrane filter of 4.5-μm effective pore size.

Division, at Pleasanton, California. There Drs. S. C. Furman and H. W. Alter directed the production of Nuclepore filters, the new and present commercial name.

A MODEL TRANSITION

In the 1960s, including the days leading up to Nuclepore filters, industrial managers of scientific research gave intense attention to how best to use new technical ideas and results for practical applications: In short, to answering the question, How do we go from discovery to product? There were many models, but few shed much light on how best to manage transitions from science to technology. In this case, however, the startlingly rapid transition from holes etched in the laboratory, to pilot plant, to commercial reality was exceptional not only in its speed, but in its closely matching the managerial strategy under which it was spawned!

Our segment of the Research Laboratory was managed in the years leading up to 1960 by an impressive, controversial, and forceful character, John Herbert Hollomon, who later became president of the University of Oklahoma. Herb produced strong reactions from those around him—admiration and enthusiasm from his associates and those for whom he was an advocate, and resentment among rival managers. He was a strong supporter of basic science in industry, and had evolved a scheme in which his department, erroneously called Metallurgy and Ceramics, was subdivided into four science groups, plus two applied groups whose jobs included taking new ideas from the science sections and transforming them into new products or services. People in all six groups were selected to do what they most enjoyed: Scientists could do science, engineers could do practical things, and the good of General Electric was served. Westbrook and I (1994) referred to Hollomon as a "wise, vigorous, effective advocate of the relevance and value of scientific research in industry."

And Hollomon's two-step process worked perfectly here. Price, Walker, Symes (who did most of the filter preparation), and I handed over filter development to Bruch at the Laboratory. He improved methods, produced filters to allow Seal and others further to test their usefulness in cell separation, and shortly relayed the responsibility for commercial production to the group at Vallicitos.

In addition to being innovative, Hollomon was versatile. He was the person who recruited me to come to the Research Lab from a position as an Assistant Professor at M.I.T. – with his format for doing science in industry being one major attraction. After I accepted G.E.'s offer, but before leaving M.I.T., Hollomon came to talk at the Faculty Club, a talk I attended with anticipation. He began with something like, "Well, we know about as much science as we need. Now is the time to go out and use it." I shortly understood this shocking reversal when I learned that since I had last seen him, he had taken a new job as the head of G.E.'s General Engineering Laboratory.

HOW ARE HOLES MADE COMMERCIALLY?

Using fission fragments from ^{252}Cf in an evacuated bell jar is not an efficient process for mass-producing filters (see Figure 2-3). How were Nuclepore filters first made commercially, and how has processing evolved? Originally, plastic film was continuously wound from one spool through a port into a nuclear reactor, past a plate with a thin layer of ^{235}U and back out of the reactor onto a second spool. The reactor neutrons induce fission of ^{235}U and the fission fragments enter the plastic in numbers that depend on the geometry, the amount of ^{235}U, the neutron flux, and the speed of the tape. For etching, the tape is threaded through sodium hydroxide in a heated tank, then a rinsing bath, and is dried before being rewound or cut to useful sizes.

A collimating honeycomb separates the fission plate from the film during the reactor exposure and insures that the preponderance of the fission fragments arrive close enough to normal to the film that they cross it, rather than being stopped. Instead of being perfectly aligned, hole orientations are grouped within 30° to 35° of normal to the sheets. This variability has a beneficial effect in minimizing coincidences where holes from adjacent tracks overlap to produce oversized holes, which become more abundant as the porosity increases. In Figure 2-1 some cases of overlapping diamond-shaped holes are visible. But where tracks are not parallel, most such overlaps don't continue all the way through the filter, and hence oversize particles or cells do not leak through.

Nuclepore was first a General Electric product, then the business was separated into a new independent company—Nuclepore Corporation of Pleasanton, California, partly owned by G.E., and later sold. It now exists as part of Corning Incorporated Separations Division, with headquarters in Cambridge, Massachusetts.

Methods have changed over the years. Now heavy ions from an accelerator (fake fission fragments) are used by Nuclepore. Such irradiation gives the option of making nicely aligned, perpendicular incidence holes, which are preferred for ease of observation in a microscope when the particles that are caught on a filter are to be examined *in situ* (see Figure 2-4). When high porosity is wanted, some jiggling of the beam or plastic sheets is used to minimize overlapping etched tracks. This jiggling can be done by accelerator optics or by the classical means of producing assorted angles of incidence of accelerator particles by interposing a thin metal foil in the beam, so that so-called Rutherford scattering of the accelerator particles by the atoms in the foil spreads the angles of incidence. A commercial virtue of using accelerated particles is that, unlike most fission fragments, no radioactivity need be involved. With reactor-based irradiations there is usually some slight, low-level residual radioactivity – of

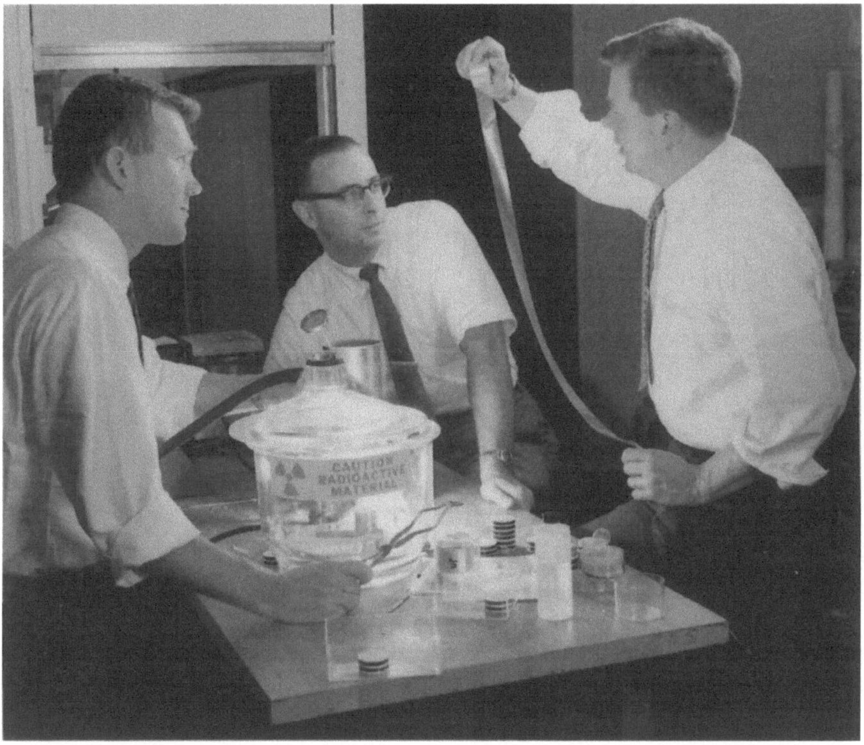

Figure 2-3. P. Buford Price, the author, and Robert M. Walker displaying tabletop apparatus for irradiating filter material. If Walker looks as though he is concentrating and stressed, try holding your arm in that position for an hour while a General Electric photographer positions and lights the subjects thoroughly (very thoroughly).

little real importance, but nevertheless a public relations hazard. A real merit of accelerator particles is that they make identical tracks since they are all the same nuclear species and energy – unlike fission fragments, which have wide ranges of mass and energy (Hyde, 1964).

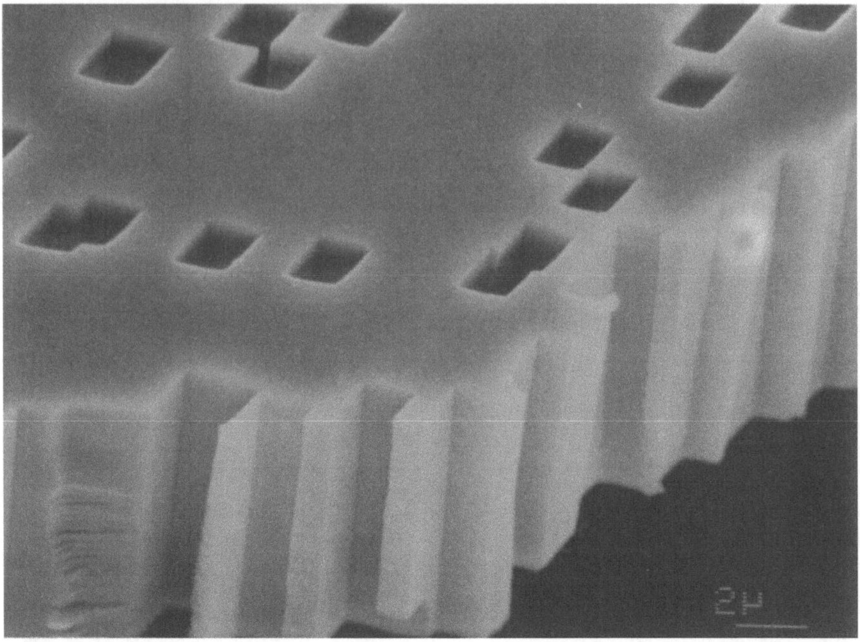

Figure 2-4. A cut-away section of a mica filter made from aligned ions from an accelerator. A few overlaps can be seen (Spohr, 1980).

Subtleties abound in track etching of polymers (FPW75). Often cone angles can be lowered by aging in air or exposure to ultraviolet light – ways of producing holes of nearly constant bore. The composition of the etch can also affect hole uniformity markedly. Sometimes, because of molecular alignment, plastics have lower etching rates normal to the sheets than in their planes, allowing larger holes to be etched than one would otherwise expect would be possible before the whole sheet is etched away. Holes from 10 nm to 18 μm are available commercially.

Nuclepore, which had a gross income of more than $10,000,000 in 1988, has continued to grow; and with the expiration of the original patents, it has experienced competition from new entries in the track-etching business. Poretics in Livermore, California, was founded by former Nuclepore people. This firm still uses reactor irradiation for the track-forming step. Other businesses exist at the heavy ion accelerator (Research Center for Applied Nuclear Physics) at Dubna in Russia, in what was East Germany, and at Whatman S.A. in Belgium. At Dubna they estimate the annual worldwide demand for etched-track filters to be about 500,000 m^2 (Kuznetsov *et al.*, 1995) – of which 40% is for medical uses; 40% for production of clean high-purity components in the electronics industry; and 20% for analytical, environmental, air purification, and various laboratory applications. At 1993 prices this estimate is equivalent to $50 million per year.

WHERE DO HOLES LEAD?

As just noted, filters with well-defined holes have many uses. One of the more obvious applications is to strain particles from liquids with unit efficiency for objects that are too large to fit through the holes. Figure 2-5 (Lal, 1977) is a case

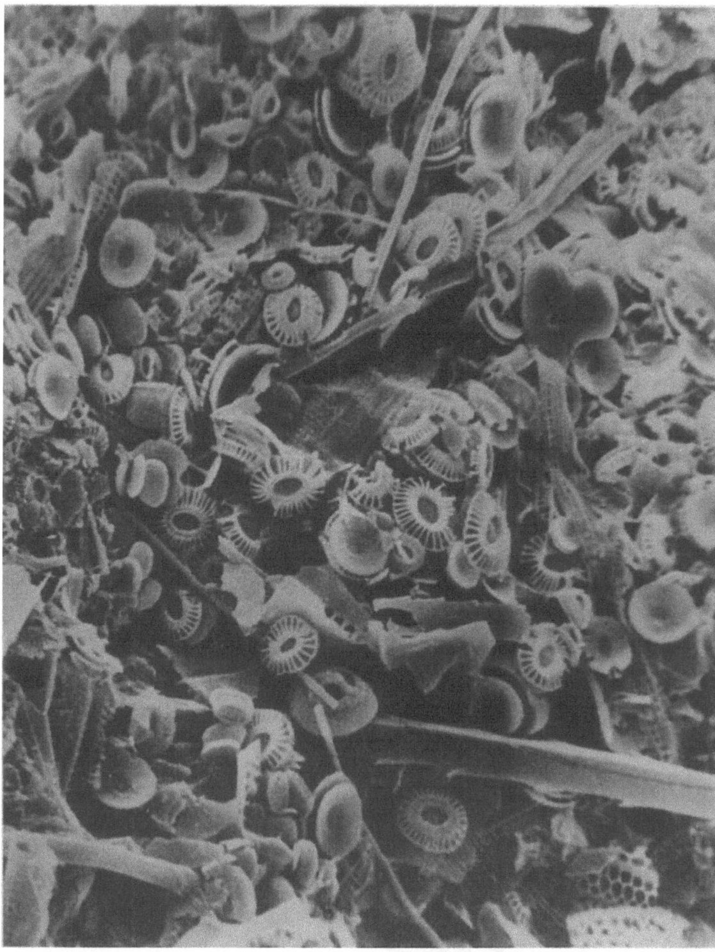

Figure 2-5. Scanning electron micrograph of suspended material collected by filtration of a large volume of Pacific surface water near 40°N, 160°E. Particles are a variety of microscopic marine plants, mostly coccolithophores and some diatoms and animals. The most abundant coccoliths, 3 – 4 μm in diameter with rays, belong to the species *Emiliania huxleyi*. Hard skeletal parts of biogenic particles fall from surface to deep waters. At all depths, they dominate the suspended phases and their dissolution and attrition, in course of settling, has an important effect on the chemistry of seawater. Photo by V. G. Shah, Physical Research Laboratory, Ahmedabad, India (Lal, 1977).

where diverse minute marine plants and animals from ocean water were sieved out in such abundance that one can barely identify an occasional underlying etched-track hole. (Over the years I've seen a good many cover pictures in the journals *Science* and *Nature*, where subjects were photographed on backgrounds of Nuclepore filters.) Oceanographers and limnologists routinely use filters to extract solids from water to reveal its particulate content.

Sampling particles from air is also a major use of filters. As in the filtering of liquids, all oversize particles are caught; but so too in this case are many of the smaller ones. Figure 2-6 shows the efficiency for collecting particles using, for example, holes of 2.5-μm radii (Spurny *et al.*, 1969). For sizes less than, but close to, the hole size, most particles are caught on the filter by impaction, as the air flow toward holes must curve to enter the holes and the larger particles fail to follow the air flow perfectly. At the other extreme, the very fine particles diffuse

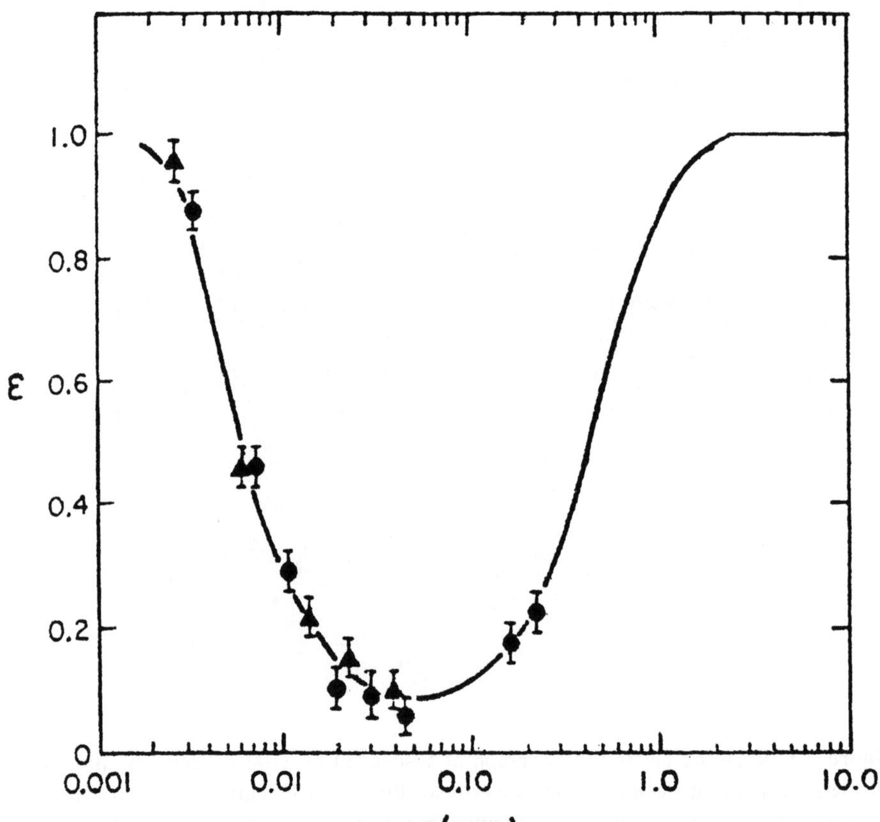

Figure 2-6. Filter efficiency in capturing aerosols. The computed curve is compared with results for selenium particles (filled circles) and pyrophosphoric acid aerosols (solid triangles) for hole radii of 2.5 μm and a 5-cm/sec face velocity of air flow (Spurny *et al.*, 1969).

rapidly and attach to the surface or to the interior sides of pores. Only an intermediate size penetrates successfully, the critical size differing with the flow velocity and the hole size. Hence, by using a series of filters of different hole sizes and/or flow velocities, virtually all airborne particles can be efficiently collected for examination.

As one example, Figure 2-7 shows a common variety of aerosol particles that are found in marine air, NaCl from evaporated sea spray. All of the particles shown are small enough to have crossed through the filter had they not impacted the surface or diffused to it. Figure 2-8 shows yeast extracted from beer as a means of allowing draft beer to be canned and then stored at room temperature. Not surprisingly, our testing program during this work received many offers of assistance.

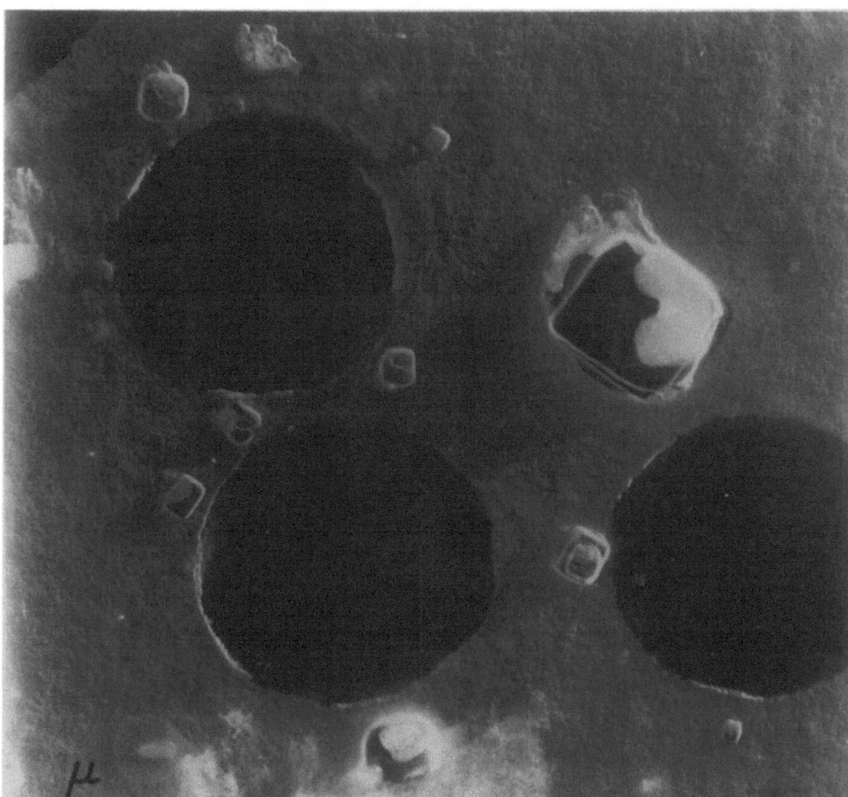

Figure 2-7. Salt particles caught from the air near Hawaii by a filter with holes of 1.8-μm diameter (photo by E. R. Frank, National Center for Atmospheric Research [NCAR]).

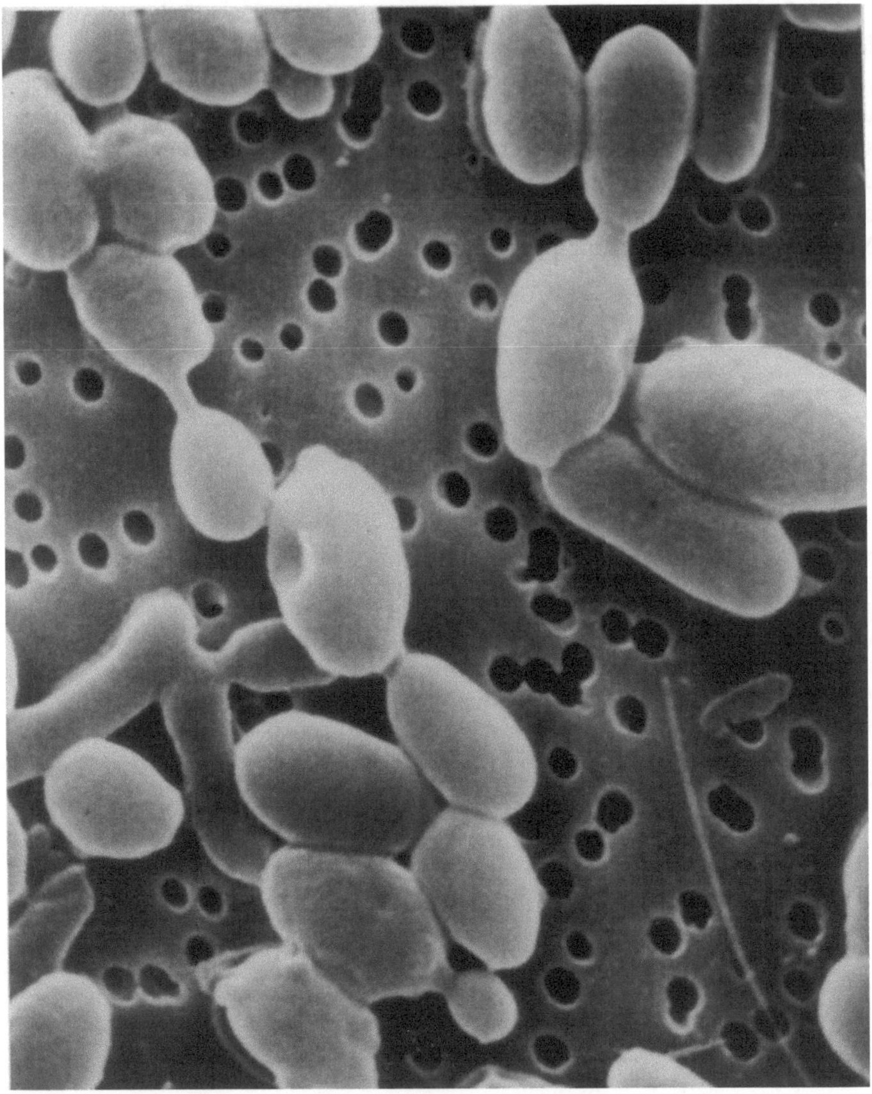

Figure 2-8. Yeast extracted from beer using Nuclepore filters (photo by E.R. Frank of NCAR).

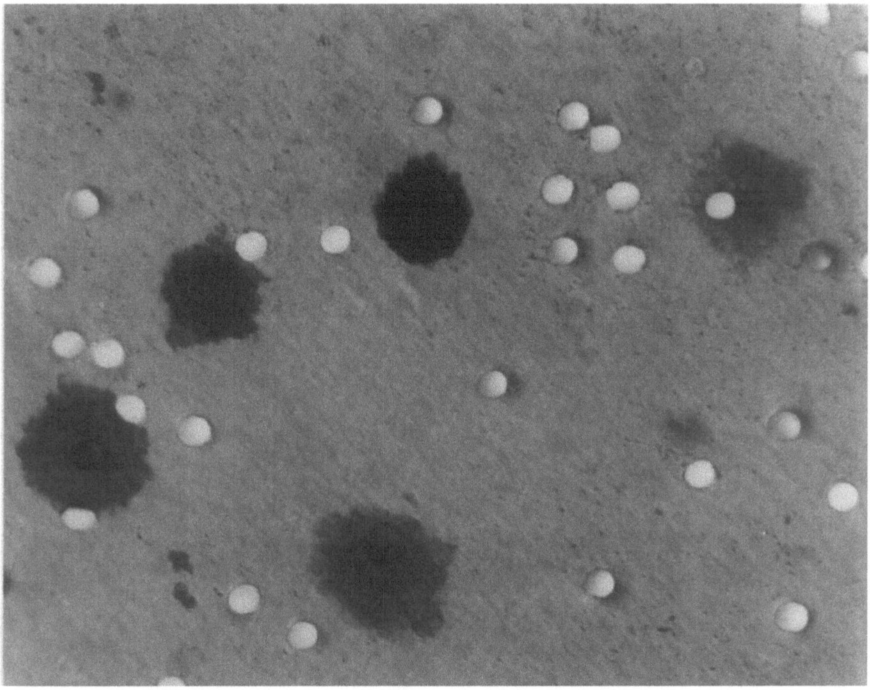

Figure 2-9. Filters of 5-μm hole size are used to filter cancer cells out of blood. The large, dark blobs are cancer cells that can be filtered out of a large quantity of blood by virtue of being larger and more rigid than most normal blood cells (Fleischer *et al.*, 1965).

Dr. Seal's idea for sieving free-floating cancer cells in blood was made into a successful technique for isolating such cells (Seal 1964) by using filters with large enough holes (4.5-μm diameter) that most blood cells can squirm through, leaving only the occasional larger, more rigid ones, as shown in Figure 2-9. Seal showed that when malignant cells are present they can be separated and recognized. The next vital question is whether such cells are present in the blood early enough during the disease that locating them will allow their source to be identified and treated effectively. To that purpose Song *et al.* (1971) examined blood samples from cancer patients and found that more than two-thirds of those with cancer of the breast, colon, or rectum had detectable cancer cells in the blood in cases where the malignancy had not metastasized, in contrast to less than 1 in 12 having such free-floating cells once the cancer had spread to the lymph nodes. Although unexplained, the direction of the effect is encouraging in that a malignancy is detectable early, when treatment could be most effective.

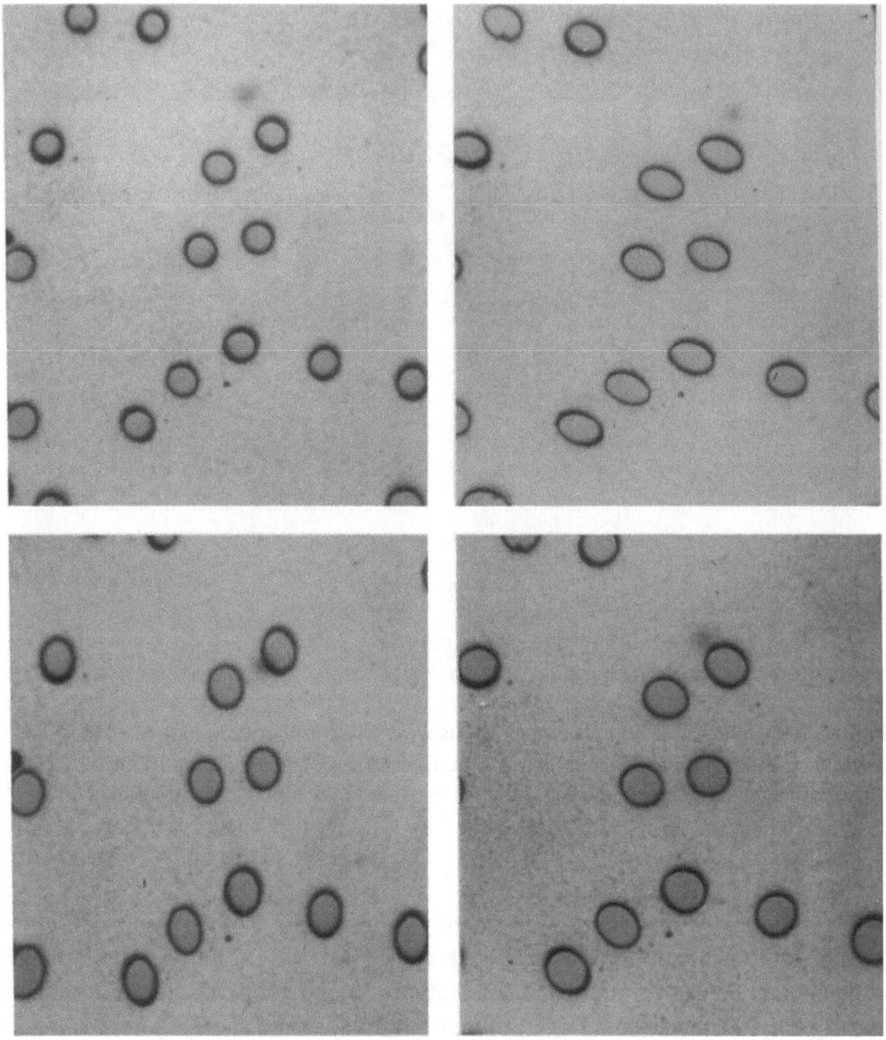

Figure 2-10. Stretchy filter made from a sheet of silicone-polycarbonate copolymer by etching fission fragment tracks. Shown in nearly relaxed condition, horizontally stretched, vertically stretched, and biaxially stretched. The hole diameters in the relaxed state are 14 μm (Fleischer *et al.*, 1972).

Another interesting type of filter, but one that is not yet a commercial product, is made of stretchy plastic, so that holes can be shaped by straining a membrane, as shown for example, in Figure 2-10. If a filter is placed on a camera iris, the pores can be continuously adjusted in size, including the possibility of enlarging them enough that a clogged filter might be unplugged. As sketched in Figure 2-11, the elasticity makes it possible to make small holes in films whose thickness exceeds the range of the particles. As sketched in part a, the tracks can be etched through less than half of the thickness. But by stretching as in part b, irradiating, relaxing, and then etching, the thicker sheet can be made into a filter. This flexibility raises the further possibility of twisting a sample, irradiating, relaxing, and etching to produce corkscrew-shaped holes. Presumably left- and right-handed helical worms could be separated. Such a separation is perhaps not a pressing need of society; however, its use would definitely be a new twist.

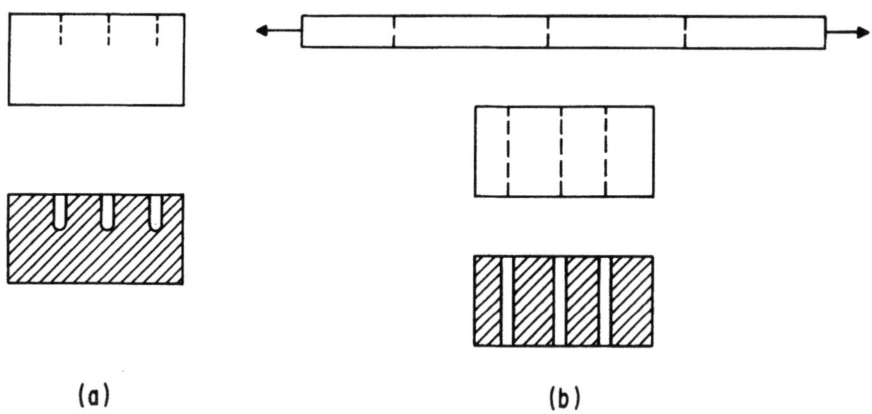

Figure 2-11. Etching of holes through thick sheets of material. In sequence a, the fission fragments do not cross the sheet. In sequence b, the sheet is stretched before irradiation; the tracks cross the sheet, and the sheet is relaxed and then etched to produce holes through the thick sheet (Fleischer *et al.*, 1972).

SINGLE HOLES—PARTICLE AND HOLE CHARACTERIZATION

A more frugal use of etched tracks is to provide single holes for particle counters and particle characterization. And we will see that the technique provides insight into the character of track damage and of etched holes.

The principle and a schematic of the DeBlois Bean Counter, named after its inventors, DeBlois and Bean (1970), is sketched in Figure 2-12. A membrane with a single hole is a barrier in a conducting solution, separating the two electrodes of an electrolytic cell; the conductance is then monitored with an oscilloscope. Once an insulating particle enters the hole, it crowds the current path and therefore increases the resistance by an amount that DeBlois and Bean showed is proportional to d^3/D^4, where d and D are, respectively, the particle and the hole diameters (for d much less than D). Thus, the pulse height gives the volume of the particle, and the pulse length inversely measures the particle mobility, which increases with any net charge that it carries.

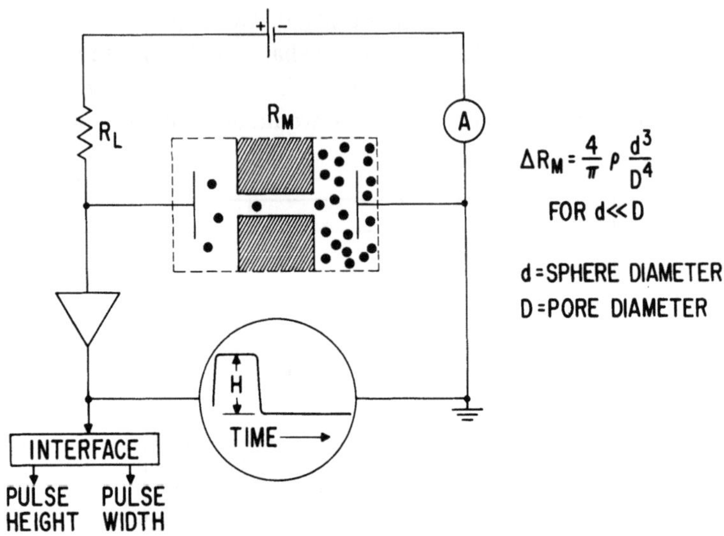

Figure 2-12. Principle of the DeBlois bean counter. A membrane with a single hole separates two electrodes of an electrochemical cell. When an insulating particle moves through the hole, the resistive pulse is proportional to the volume of the particle and the inverse fourth power of the hole diameter (DeBlois and Bean, 1970).

How Are Single Holes Made?

The original technique used by DeBlois – a superbly adept experimentalist – for producing a membrane with a single hole was to irradiate with an exposure that was designed to produce only a few tracks in a small area. He microscopically located the holes in the etched membrane and stopped off all but one. More recently at the GSI (the heavy ion accelerator at Darmstadt, Germany) engineers have learned how to gate the beam so as to irradiate with a single ion (Roggenkamp *et al.*, 1981).

A resistive pulse device (made with a drilled hole) was created by and named after W. H. Coulter (1950, 1953), and it is available commercially. Using an etched track allows good size resolution for smaller particle sizes. As an example, the sperm pictured at top of Figure 2-13 have a distinctive shape that is strikingly reflected in the resistive pulse shown below (DeBlois, 1978). A high pulse records the passage of the head (volume = 5.5 ± 0.5 μm^3) and a long low pulse the subsequent passage of the narrow tail (diameter = 0.26 ± 0.01 μm). Demonstrating the agility of an otter, a second sperm passed through tail first.

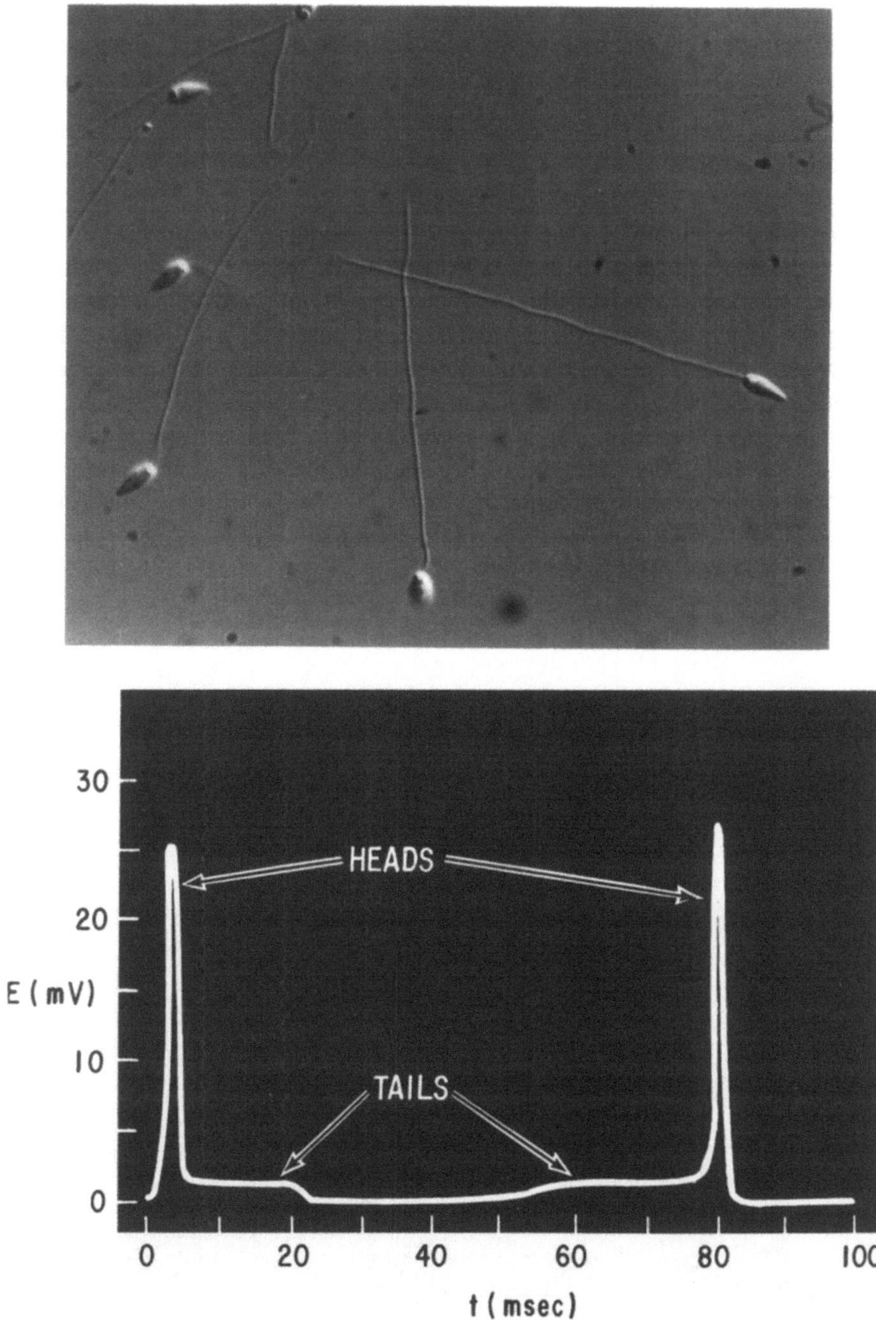

Figure 2-13. As sea urchin sperm (top) move through a hole in a membrane, the resistive pulses (bottom) show one entering head-first and the second tail-first (photo from C. P. Bean, pulses from R. W. DeBlois).

As a further example, the resistive pulse technique is effective in sizing viruses (DeBlois and Bean, 1970), using as controls spheres of polystyrene of known diameters (in this case with a nominal value of 91 nm). Comparison of the two pulses in Figure 2-14 shows that the T-2 virus is 115 nm in diameter.

Hole Uniformity

The resistive technique also has great power in measuring hole dimensions. Although there is some noise in the oscilloscope traces, the two pulses in Figure 2-15 are remarkably similar. The local fluctuations reflect diameter variations along the length of this particular rather rough, tapered pore. Because of the $(\text{diameter})^{-4}$ relation noted earlier, the observed ±6% fluctuations from a smooth variation in the pulses imply roughness of ±1.5% in the bore of the hole. The 120% increase to the peak, represents a cone angle of 5° for a 10-μm-thick sheet with a 5-μm hole. The traces are highly sensitive measures of hole uniformity and can be used for quality control by firms that produce filters. Some of the additional noise in the pulse profiles results from particles traversing the holes at varying distances from the center line.

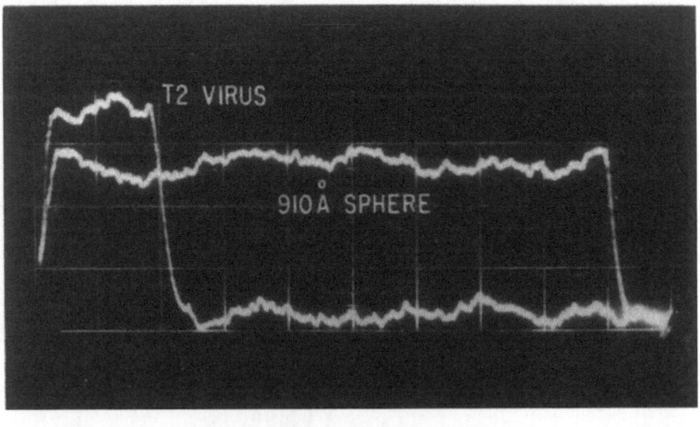

Figure 2-14. Oscilloscope traces resulting from the passage of a polystyrene sphere and a T-2 virus through the hole in the counter. Individual particles may be identified by their size (given by the pulse height) and charge (inferred from the transit time) (DeBlois and Bean, 1970).

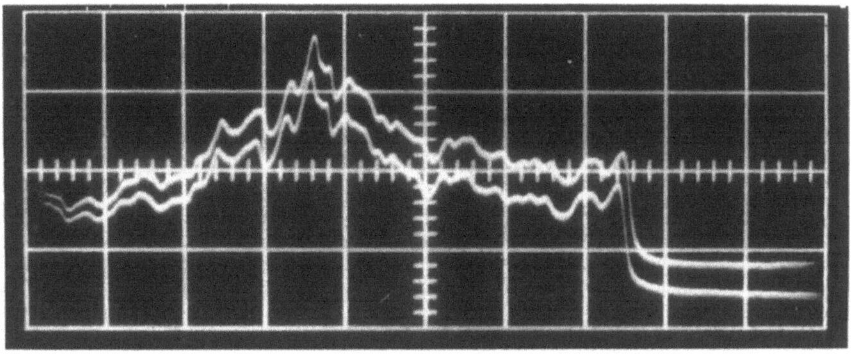

Figure 2-15. Resistive pulses as two nearly identical calibration particles transit the same hole under an applied field. The ripples are duplicated, showing they are real, small-scale (±2%) diameter variations. The increase toward the center represents a narrowing of the hole and measures the taper (or cone angle) of the etched track (D. C. Golibersuch and R. L. Fleischer, unpublished).

TRACK BASICS FROM ETCHING FILTERS

A variation on the experiment sketched in Figure 2-12 reveals basic information about tracks – their sizes and the rate of radial etching, which in turn is a measure of the intensity of damage. A membrane with a few unetched tracks is used, the electrolyte is an etchant, and the resistance versus time is measured (Bean *et al.*, 1970; Bean, 1975). From the number of tracks and the resistance through the holes an average track radius can be inferred, as shown for muscovite mica in Figure 2-16. The conductivity jumps from zero at the start to that corresponding to hole diameters of 6.6 nm at 13 sec etching time. After a hindrance to hole growth (that Bean modeled as caused by surface tension) is overcome, the increase in size with time of etch becomes linear (after about 100 sec). The behavior of polycarbonate plastic films is qualitatively similar (Bean, 1975), as seen in Figure 2-17. DeSorbo (1979) used this technique on polycarbonate plastic films to study sensitively a multitude of environmental effects, including exposure to air, oxygen, ultraviolet light, nitrogen, carbon dioxide, thermal treatment, aging time, and electron irradiation. This information is of great value for reliably preparing etched-track filters.

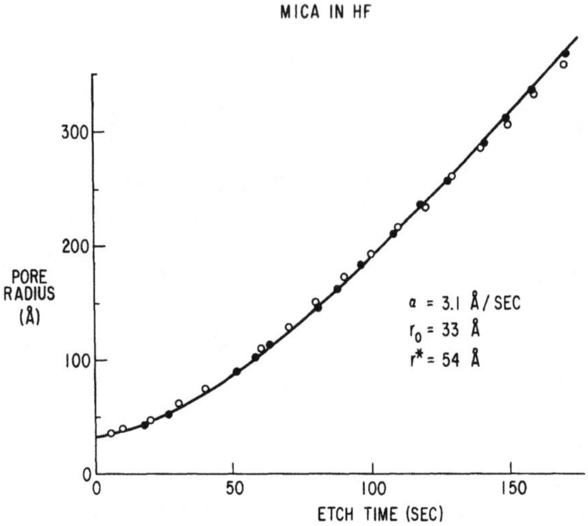

Figure 2-16. Growth of holes as measured by electrical conductivity through pores. The inferred radius as a function of time is given for two irradiated muscovite mica samples, each of thickness 4.2 μm, etched in 34% HF at 25°C. The solid curve is a theoretical result that includes the effect of surface energy in slowing the initial radial etching. The derived original radius when the etchant first penetrates is 33 Å (Bean *et al.*, 1970).

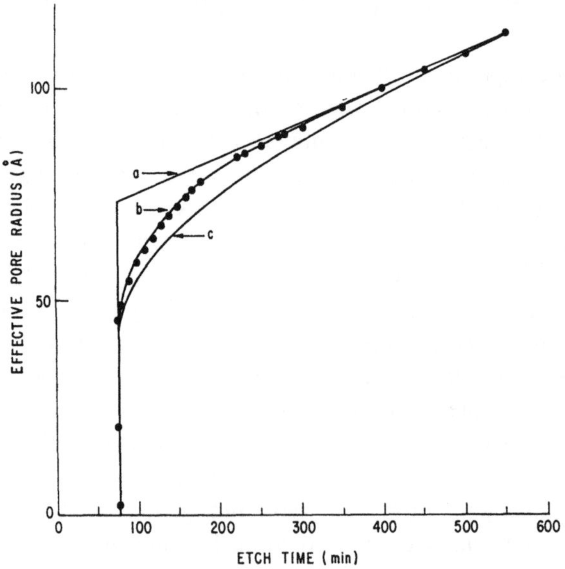

Figure 2-17. Effective pore radius as a function of etching time for a single pore in polycarbonate. The radius is inferred from electrical measurements of pore resistivity during the etching process. The solid curves are for various assumed radial dependencies of decreased activation energy for etchant attack. Curve a is a square well of depth 0.20 eV and radius 71 Å. Curve b is an arbitrary best fitting function, $0.20(1-(r/97Å)^{1/3})$eV and curve c is 0.20 exp (–r/27Å)eV (Bean, 1975).

A WIDER LANDSCAPE

Figure 2-18 is a simulated landscape – an etched-track filter decorated with polystyrene spheres. It represents the new unpredicted technological pathways that holes produced by track etching have opened. Tracks led to filters, filters in turn to cleaner semiconductors, efficient air and water samplers for environmental monitoring and oceanographic studies, a tool for virus identification and for fundamental track measurements, and a wide variety of other uses, many of which are described in references in the bibliography (FPW75, S90).

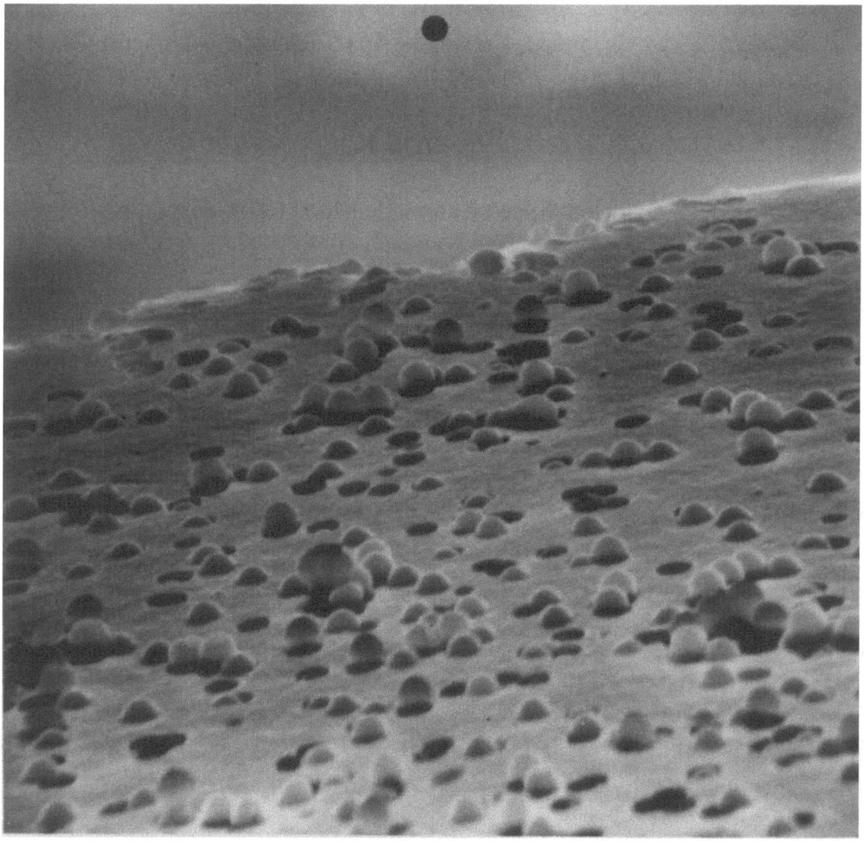

Figure 2-18. Etched track filter with polystyrene spheres (supplied by C.P. Bean).

FURTHER READING ON HOLES AND THEIR USES

FAFPW72 *Particle Track Etching* (1972), R. L. Fleischer, H. W. Alter, S. C. Furman, P. B. Price, and R. M. Walker, Science **178**, 255 – 263.

FPW75 Chapter 10 of *Nuclear Tracks in Solids*, R. L. Fleischer, P. B. Price, and R. M. Walker, Univ. of California Press, Berkeley, 1975, pages 562 to 595.

S90 Chapters 7 and 8 in *Ion Tracks and Microtechnology*, R. Spohr, Vieweg, Braunschweig, 1990, pages 183 to 329.

REFERENCES

C. P. Bean (1975), "Electrical Measurements of Track Diameter," pages 11–16 in FPW75.

C. P. Bean, M. V. Doyle, and G. Entine (1970), "Etching of Submicron Pores in Irradiated Mica," *J. Appl. Phys.* **41**, 1454–1459.

W. H. Coulter (1953), U.S. Patent No. 2,656,509.

W. H. Coulter (1956), "High Speed Automatic Blood Cell Counter and Cell Size Analyzer," *Natl. Electron. Conf. Proc.* **12**, 1034–1042.

R. W. DeBlois (1978), "Analysis of Morphology of Sperm Cells Using a Resistive-Pulse Technique," *Biophys J.* **24**, 149.

R. W. DeBlois and C. P. Bean (1970), "Counting and Sizing of Submicron Particles by the Resistive Pulse Technique," *Rev. Sci. Instr.* **41**, 909–916.

W. DeSorbo (1979), "Ultraviolet Effects and Aging Effects on Etching Characteristics of Fission Tracks in Polycarbonate Film," *Nuclear Tracks* **3**, 13–32.

R. L. Fleischer and P. B. Price (1963), "Tracks of Charged Particles in High Polymers," *Science* **140**, 1221–1222.

R. L. Fleischer, P. B. Price, and R. M. Walker (1963), "Method of Forming Fine Holes of Near Atomic Dimensions," *Rev. Sci. Instr.* **34**, 510–512.

R. L. Fleischer, P. B. Price, and E. M. Symes (1964), "A Novel Filter for Biological Studies," *Science* **143**, 249–250.

R. L. Fleischer, P. B. Price, and R. M. Walker (1965), "Tracks of Charged Particles in Solids," *Science* **149**, 383–393.

R. L. Fleischer, J. R. M. Viertl, and P. B. Price (1972), "Biological Filters with Continuously Adjustable Hole Size," *Rev. Sci. Instr.* **43**, 1708–1709.

E. K. Hyde (1964), *The Nuclear Properties of the Heavy Elements, III. Fission Phenomena*, Englewood Cliffs, NJ, Prentice-Hall.

V. I. Kuznetsov, L. V. Kuznetsov, and V. D. Shestakov (1995), "Track Membranes of the Third Generation, (TMG-3): Their Properties and Industrial Application," *Nucl. Instr. Methods.* **B105**, 250–253.

D. Lal (1977), "The Ocean Microcosm of Particles," *Science* **198**, 997–1009.

H. G. Roggenkamp, H. Kiesewetter, R. Spohr, U. Dauer, and L. C. Busch (1981), "Production of Single Pore Membranes for the Measurement of Red Blood Cell Deformability," *Biomed. Technol.* **26**, 167–169.

S. Seal (1964), "A Sieve for the Isolation of Cancer Cells and Other Large Cells from the Blood," *Cancer* **17**, 637–643.

J. Song, P. From, W. Morrissey, and J. Sams (1971), "Circulating Cancer Cells: Pre- and Post-Chemotherapy Observations," *Cancer* **28**, 553–561.

R. Spohr (1980), "Nuclear Track Irradiations at GSI," *Nuclear Tracks* **4**, 101–108.

K. R. Spurny, J. P. Lodge, Jr., E. R. Frank, and D. C. Sheesley (1969), "Aerosol Filtration by Means of Nuclepore Filters," *Environ. Sci. and Technol.* **3**, 453–464.

C. G. Suits and A. M. Bueche (1961), "Cases of Research in a Diversified Company," in *Applied Science and Technological Progress* (report to the Committee on Science and Astronautics, U.S. House of Representatives by the National Academy of Sciences), Washington, D.C., U. S. Gov. Printing Office, June, pp 297–346.

J. H. Westbrook and R. L. Fleischer ed. (1994), *Intermetallic Compounds—Principles and Practice*, Chichester, UK, J. Wiley Ltd.

Chapter 3
RADON—A HAZARD AND A HELP

They are ill discoverers that think there is no land,
when they can see nothing but sea.
Francis Bacon, 1605, in *The Advancement of Learning*

A HAZARD

Stanley Watrous back in 1985 was having an awkward time. For several days, as he finished work at the Limerick nuclear facility in eastern Pennsylvania, he tripped the alarm during his routine radiation check at the end of the day. Yet there was no unusual loose radioactivity to be found in his work space. One morning, in a moment of inspiration, or perhaps of desperation, he checked himself when he arrived at work, and this check showed far more radiation than had caused the earlier alarms. A different alarm then went off in Watrous' head.

From the title of this chapter many readers will correctly have realized that the radioactive gas radon is the bad actor in this scene. Watrous shortly found that his home contained radon in its air at a concentration of 3,000 pico-Curies per liter (pCi/l), about 3,000 times more than is usual in homes in the United States. (A useful physical feeling for what a pCi/l amounts to is one disintegration per minute per pint, an easily remembered unit that appears to be designed to unhinge enthusiasts of standard units.) From his home Watrous was carrying to work radioactive products of radon on his skin and clothes in amounts that were still obvious at the end of his work days, even though by then only a minute fraction of the original activity was still present.

A particular level of radiation can either be harmful or trivial relative to other normal exposures, but in this case the exposures were by no means minor. Watrous, his wife, and two children had lived in their house for 11 months. From the Environmental Protection Agency's assessment of the effects of radon on the lungs, the inference is that there is about a 50% probability of one of the four ultimately dying of lung cancer from this exposure.

NUCLEAR TRACKS FROM RADON

Nuclear tracks and radon are intertwined because a number of the plastic track detectors can record tracks from the alpha particles that are emitted by radon and two of its "prompt daughters." In this chapter we are primarily interested in the most common type of radon, that of mass 222: ^{222}Rn. Its immediate parent is radium (^{226}Ra), and its earliest antecedent is ^{238}U, which is the most common variety of natural uranium atom. Hence, other things being constant, there is more radon where there's more uranium.

The prompt daughters are four radioactive lead, bismuth, and polonium atoms that are produced by a sequence of decays within about an hour on average of each ^{222}Rn decay. Two of these are polonium atoms, ^{218}Po and ^{214}Po. Because they are alpha emitters, these species are the direct troublemakers in the relation of radon to health. These atoms irradiate the surface of the lining of the lung most destructively. Alpha emitters, among the common natural radioactivity substances, are especially troubling in that they deliver the most energy in the shortest distance. An alpha particle is merely a moving helium nucleus, and a typical natural alpha particle can be stopped by a piece of paper. But within the depth that alphas do penetrate they destroy or alter a great many interatomic bonds, giving the alpha particles the potential to kill cells, or worse, to induce malignant changes.

Radon, although itself alpha-active, rarely delivers much radiation to the lung directly, since most of it that is breathed in is immediately exhaled. But the daughters are metallic atoms, and therefore stick to surfaces of aerosols or directly to surfaces within the lung. The lung structure is tortuous (Figure 3-1), so that most atoms and aerosols will contact the surface, stick, and allow the prompt radon daughters to decay at locations where the alpha particles can damage cells.

MEASURING RADON DOSES

Track detectors for measuring radon are the dosimeters of choice for a number of uses, first because they are passive and thus need no power supply, and second because they measure total radon exposure, averaging the radon concentration over rather long periods of time, such as a week or a year. For short-term measurements electronic and other methods are preferred. But, as will shortly be clear, there are many cases where long-term radon values are needed or at least highly preferable.

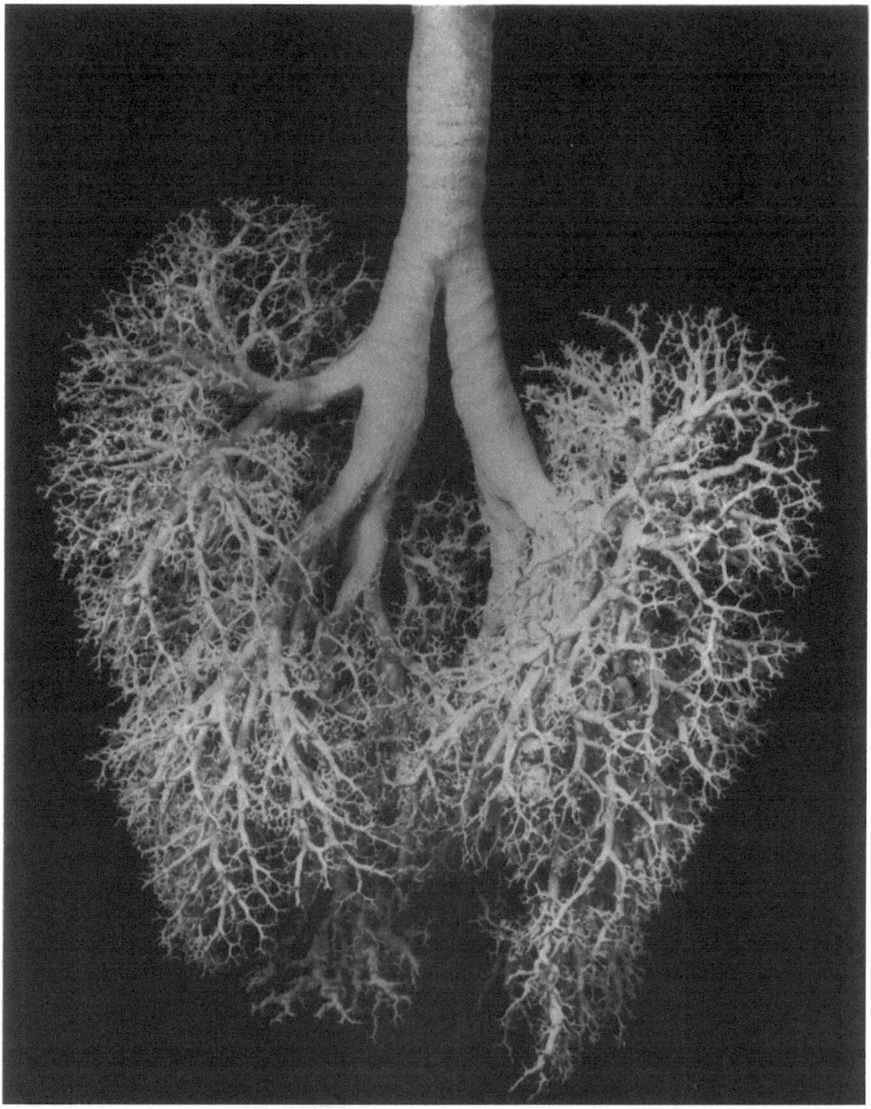

Figure 3-1. A cast of a bronchial tree of a rabbit. Deposited radioactive alpha emitters trapped in the lung irradiate the lining and can induce lung cancer.

The most obvious such case is where a homeowner wants to know whether his house is safe, or alternatively needs changes. Figure 3-2 shows that radon has multiple routes into a house, and radon also can escape through various openings, such as windows, doors, chimneys, and permeable walls. Since use of doors and windows and the heating and/or cooling systems are randomly, diurnally, weekly, and/or seasonally variable, radon will vary widely and uncalculably with home use, weather, and time of year. Radon typically causes health problems over 20 to 30 years' time; hence a one-year average is a reasonable, representative time to average over, short relative to the time to develop lung cancer, long relative to usual fluctuations. Track etching is the established long-term method that is used commercially.

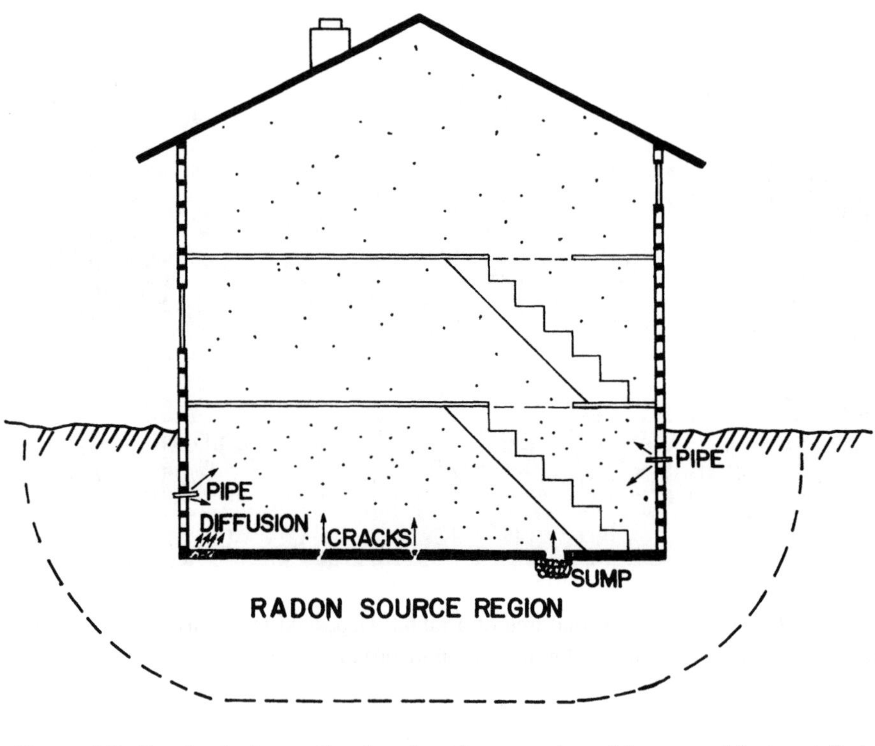

Figure 3-2. Sketch of a house showing the primary region of the ground that contributes to indoor radioactivity by releasing radon that then can diffuse to the foundation and enter the basement by various paths.

Some workers have simply exposed a bare piece of alpha-particle-detecting plastic in room air, and later counted tracks to determine radon. However, since both radon and its daughters alpha-decay to form tracks, the result is an unspecified composite of radon and of radon daughters. The uncertainty comes about because the daughters deposit on the plastic or on surfaces near it and decay to form tracks in numbers that depend on many factors. Since deposition depends on diffusion, aerosol content of the air, air circulation, and the number and position of all of the surfaces in the airspace, the ratio of daughter alpha-tracks relative to radon alpha-tracks can vary widely even when radon itself does not. Nor, without elaborate additional measurements, is the track count a reliable measure of the daughter content of the air.

Figure 3-3 shows how an unambiguous ^{222}Rn measurement may be done, as developed from the original Alter and Price (1968) patent by Antonio Mogro-Campero, William Ward, Richard Likes, and the author (and as is done commercially by Terradex, Inc., now owned by R.S. Landauer & Co., Glenmont, Illinois). Pictured is a cup with a controlled opening, and a rectangular piece of alpha-detecting plastic at the closed end. The membrane serves to keep out dust that may well carry unwanted alpha radioactivity, such as uranium and thorium, along with their daughters, including the prompt daughters of ^{222}Rn. If chosen properly, the membrane will also keep out all other radon isotopes except ^{222}Rn.

The combination of membrane thickness and its permeability to radon is selected to delay the radon from entering the cup for a time that is shorter than the 5.5-day mean life of ^{222}Rn and long relative to the 81-sec mean life of ^{220}Rn (called "thoron," since it derives from thorium), thoron being the most longlived of the other isotopes of radon. Because these other isotopes decay rather than entering the cup, the detector sees only alpha particles from the ^{222}Rn that does enter and the alphas of its polonium daughters that are produced after it enters (Ward *et al.*, 1977; Fleischer and Mogro-Campero, 1978). Many plastics can be chosen (see lists in F88 and DI97) but garden-variety polyethylene is a superior choice. Black sheet, 15 μm thick, is sold for keeping down weeds in vegetable patches, and hence it is truly garden variety.

The other feature of the simple device in Figure 3-3 is insensitivity to a humid environment. An envelope of $CaSO_4$ absorbs water vapor (Likes *et al.*, 1979), and the chosen polyethylene membrane is unusual among plastics in being less permeable to water molecules than to radon. It therefore helps minimize moisture levels in the cup. Hopefully, moisture is only rarely a problem for indoor measurements; but it is of critical importance for measurements in the earth, where humidity typically is so high that diurnal temperature fluctuations can cause condensation of water onto an unprotected detector.

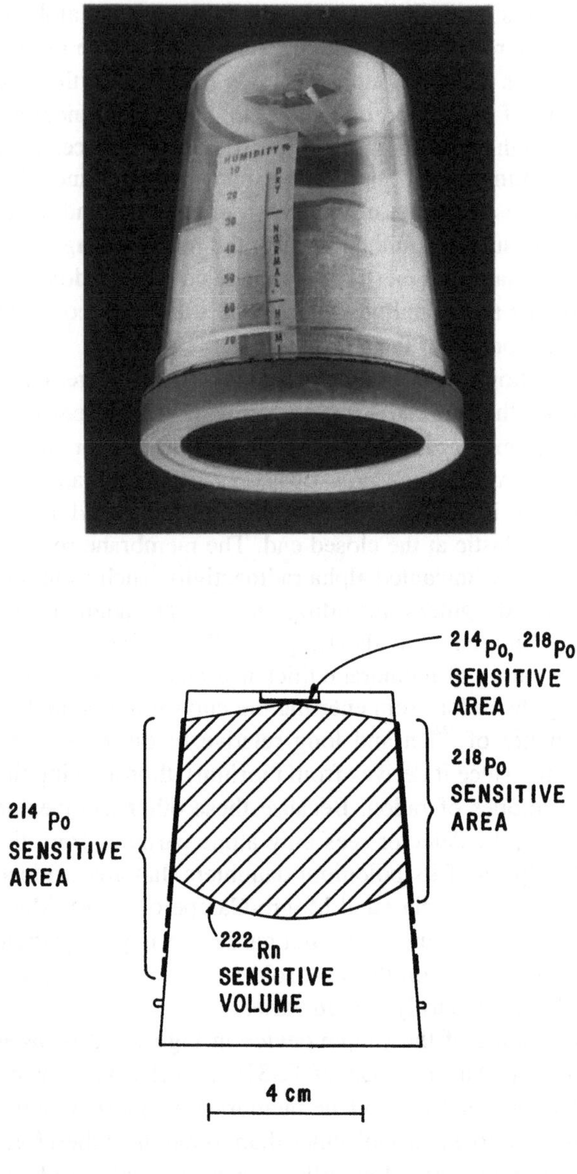

Figure 3-3. Track-etching ^{222}Rn monitoring. a) Terradex Track-Etch radon monitoring cup, with anti-^{220}Rn membrane and desiccant. A color-changing humidity gauge shows whether an acceptable humidity level is present. b) The integrating detector records the alpha decays of ^{222}Rn in the sensitive volume and those of the decay products ^{218}Po and ^{214}Po, which attach to the surfaces. This diagram applies to a detector with unit efficiency for recording alpha particles. To avoid confusion, the volumes from which ^{218}Po and ^{214}Po in the air can be recorded are not given (F88).

COMMERCIAL DEVELOPMENTS—URANIUM EXPLORATION

Radon work using tracks was started with the long-standing idea that the presence of radon is evidence of where uranium is located. Since radon derives from uranium, it will be more abundant near uranium; and being a nonreactive ("noble") gas, it can move freely through porosity to deliver the message that uranium is nearby.

The part of General Electric that was assigned responsibility for commercial use of etched tracks was Vallecitos Atomic Laboratory, part of the nuclear energy business. Its energetic and innovative manager, Dr. H. Ward Alter, wanted to insure that adequate supplies of uranium ore were available to power nuclear reactors economically. On a 1967 visit to the Research Laboratory he raised the question of whether we could use tracks to search for uranium, noting that radon emitted alpha particles and asking, "Can we measure radon?" Buford Price happened to be in the audience, and he was well aware that alpha particles produce tracks in cellulose nitrate plastics (Fleischer *et al.*, 1964). This meeting led to the initial tracks-and-radon patent (Alter and Price, 1968) for a set of devices that evolved into the improved 1979 detector shown in Figure 3-3. Surveys using track etching were done as early as 1970 (Weidenbaum).

In practice, uranium surveys are done over a grid of locations with emplacements like that sketched in Figure 3-4 (Fleischer and Mogro-Campero, 1985). A sewer pipe of polyvinylchloride keeps a hole from collapsing, a detector cup such as that described by Figure 3-3 is placed on the soil, and the intervening space is filled by an impermeable plug to prevent atmospheric, low-radon air from significantly diluting the soil radon at the base of the hole. It is particularly important to exclude ^{220}Rn. Because it is shortlived (56-sec half-life), it moves only a short distance; and it therefore cannot deliver a message from afar.

Figure 3-5a shows an apparently successful such survey over a 100-m-deep ore body near Thoreau, New Mexico, in May and June 1978. Seven of the highest readings overlie the ore body and define its trajectory, the path of an ancient stream bed. If this sort of correlation were universal, about half of the exploratory holes drilled would reach ore, thus giving a success rate far above normal experience. But life is rarely so simple, as Figure 3-5b illustrates. A separate month-long survey at the same site gave a very different result. Although again several of the highest readings are over the ore, they do not clarify its spatial distribution; and the major grouping of high readings is transverse to the ore.

These two results show the possibilities and limitations. Remarkably, we will see that the likely explanation of the reason for such changes with time has implications for indoor radioactivity, earthquake prediction, and other types of exploration for natural resources.

Gas Flow in the Earth

Why would radon in the earth vary with time? In principle it might be released at varying rates from the minerals in which it is produced by uranium (and radium) decay—released as a result of, for example, varying amounts of moisture being present. Moisture effects can occur, but not in as systematic a way or on as large a scale as the changes of interest.

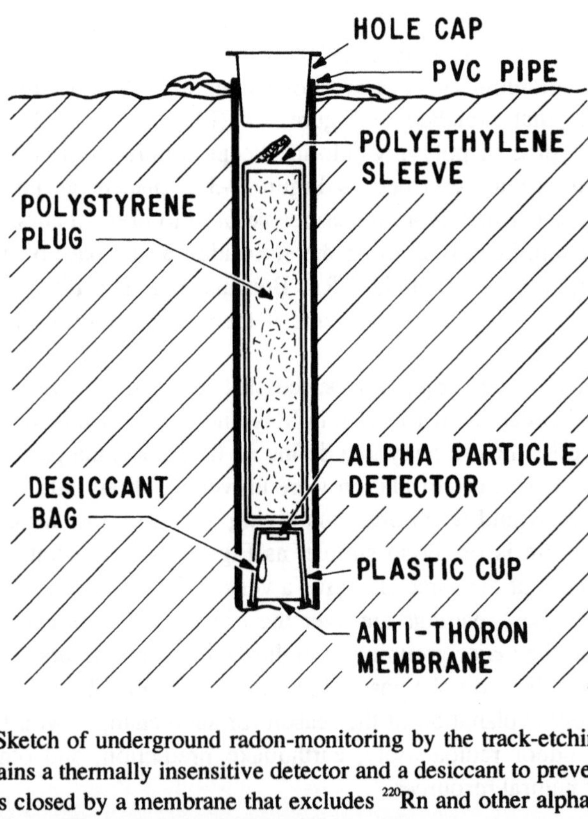

Figure 3-4. Sketch of underground radon-monitoring by the track-etching technique. A cup that contains a thermally insensitive detector and a desiccant to prevent condensation of moisture is closed by a membrane that excludes ^{220}Rn and other alpha emitters except ^{222}Rn. Insulation between the cup and the surface of the ground suppresses loss of radon to the atmosphere (F88).

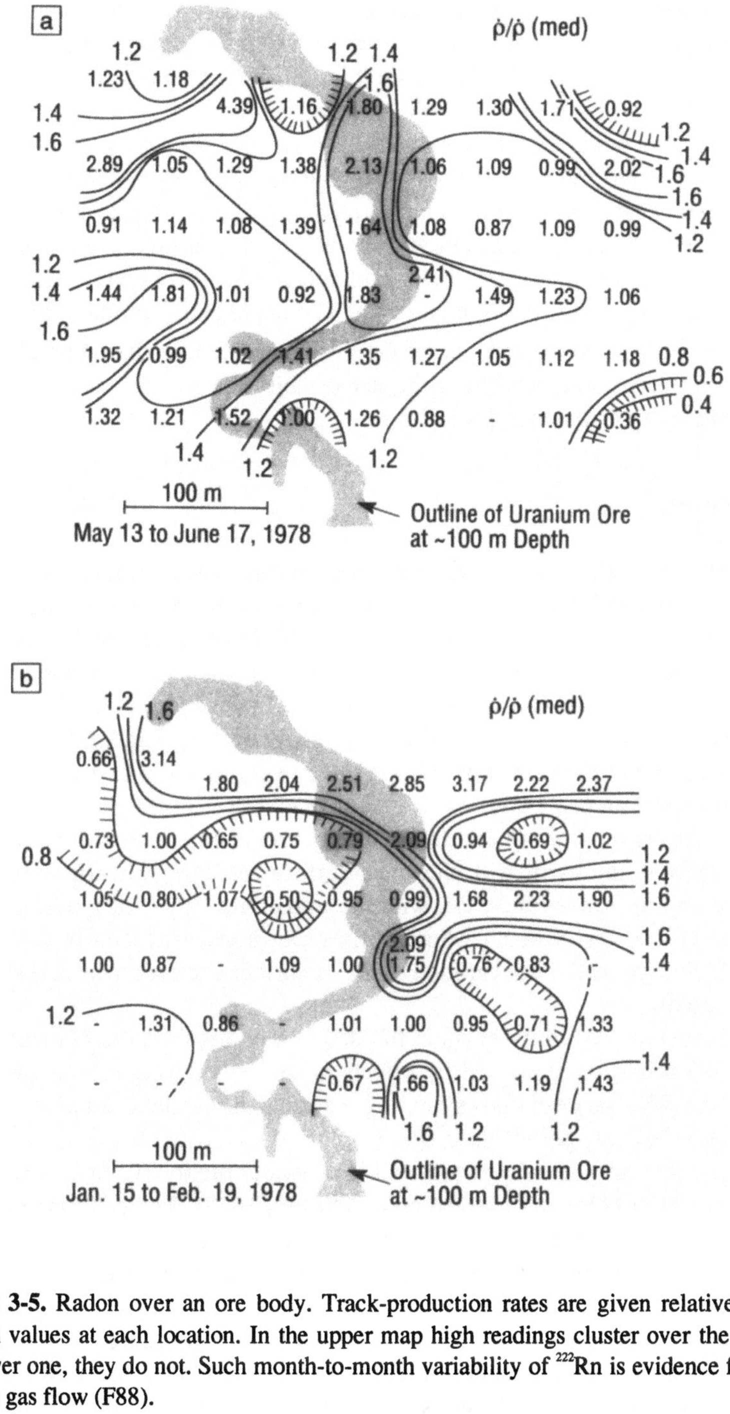

Figure 3-5. Radon over an ore body. Track-production rates are given relative to the median values at each location. In the upper map high readings cluster over the ore. In the lower one, they do not. Such month-to-month variability of ^{222}Rn is evidence for subsurface gas flow (F88).

A second explanation that fits observations far better is that motion of gas occurs through the interconnected pore spaces in soil and rock, motion that transports high-radon regions of soil gas in an easily detected way. Several causes of gas flow are possible, most of which are unproved. But the consequences for radon can be simply shown if we consider how gas flow will change the radon for the idealized case shown in Figure 3-6.

For uniform material, i.e., equal radon release into pore spaces and equal diffusion everywhere, radon will build to a constant value c_0 at which radon production and decay balance. But at the surface of the earth, radon escapes into the atmosphere, reducing the concentration to nearly zero at the air–earth interface, and producing a gradient from there. The middle curve in Figure 3-6 shows the result, a build-up from zero to c_0 at depth. The exponential variation has a characteristic decay distance $\sqrt{Dt}$ set by the diffusion rate of radon D (0.03 to 0.05 cm^2/sec in porous soil) and the mean decay time of ^{222}Rn (5.5 days). For these values the decay distance is 1.2 to 1.5 m.

A Dilemma

This distance sets a troubling limit on how deeply buried uranium ore can be and yet be detected reliably by its radon emanation. If ore has 1% uranium and typical rock 1 part per million, there is a 10,000 times enrichment above background to be recognized at the ore body. If the elevated signal is attenuated by 1/e for each 1.3 m, the 10,000× excess is reduced to background in (9.2 × 1.3 =) 12 m. Hence ore below 12 m will not be recognized by a radon measurement close to the surface—from this thinking only very shallow deposits are to be detected using radon.

The other two curves in Figure 3-6 result from another possibility, that vertical gas flow in the earth will alter near-surface readings–downflow causing a decrease and upflow an increase, with the maximum value for a uniform earth being c_0. If ore is at a depth where upflow can transport radon to the surface before it decays to the background value, a further increase will reveal where ore is present.

With such a possibility in mind the differences between the two surveys in Figure 3-6 might be explained by effects of upflows during certain times, but not always. This process makes ore at greater depth visible as a radon image at the surface, but only part of the time.

A possible source of upward and downward motion is fluid convection within the earth (Mogro-Campero and Fleischer, 1977). Typically the earth becomes hotter with depth by 20° to 40°C per km, so that the deeper fluid is buoyant and wishes to rise. Whether this effect is strong enough actually to occur at a measurable velocity depends on the properties and thickness of the permeable material and the steepness of the geothermal gradient. Measurements of ground properties show that convection might have occurred at the New Mexico site, but properties of the ground were sufficiently nonuniform that proof is lacking.

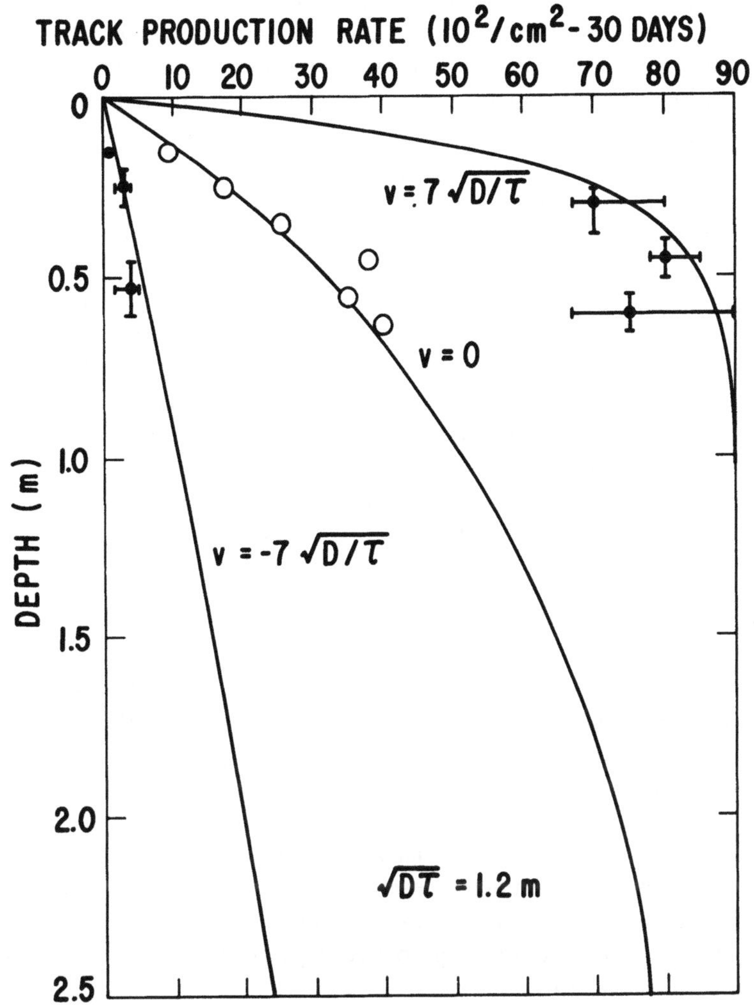

Figure 3-6 Radon concentration versus depth, calculated for zero vertical flow and flows per unit porosity of $\pm 7\sqrt{D/\tau}$, where the mean diffusion distance $\sqrt{D\tau}$ is 1.2 m and the mean diffusion time τ is 5.5 days (the mean life of ^{222}Rn). The horizontal scale is calculated to correspond to typical field values encountered at a site that we studied for several years at Thoreau, New Mexico. The open circles are observations for no vertical flow, the filled circles for specific times of up- or downflow.

The uncertainty of relying on flow might appear to disqualify an exploration method, but such is not the case. All remote exploration methods have uncertainties, but a technique may be economically useful if it increases the odds that a given drill hole will encounter ore – drilling holes for ground truth being the expensive part of the discovery process.

The 100-m-deep ore body of Figure 3-5 was in fact discovered by drilling that was guided by an earlier radon survey. It will, however, never be known whether upflow in that area was responsible for the decisive high-radon readings.

As an example of the industry's willingness and necessity to live with uncertainty, I remember a visit in Arizona to a uranium-exploration company whose aid we were seeking – loan of some equipment for drilling shallow holes. During our discussion they rolled out a map of a recent radon survey they had done and asked *me*, "Where should we drill?" Being a total novice at exploration, as they were at using the radon tool, I offered a hedged suggestion that if I were using only radon, points A, B, and C would be where I would test first. The blind leading the blind. Seeing into the earth is difficult.

Terradex Corporation

In early 1974, General Electric's radon-based exploration business was set up as a separate company, Terradex (terrestrial radon exploration), with Alter as president. Impressively, 1974 turned out to be the year that the price of uranium, which had been stable over at least the last seven years, abruptly nearly tripled. Terradex busily served the uranium industry for the next six years – with a mix of the hoped-for successes and the statistically likely failures – until exploration came to a striking downturn in the years 1980–1986. The causes were partly that newly found larger, richer ore bodies could more cheaply supply the industry, and partly that public alarm over purported hazards of radioactivity had slowed orders for nuclear power.

Many factors that make difficulties for one market contain incidental associated advantages for others. Such an effect buoyed Terradex's fortunes. Public alarm over radioactivity was paralleled by progressive technical interest in radon and health – one evidence being a rapid rise in the numbers of papers on radon in the journal *Health Physics* by a factor of seven from 1977 to 1983. And widespread public interest was aroused in 1985 by the Watrous story that was told at the beginning of this chapter. Terradex, which since 1986 has been part of R. S. Landauer and Co., continues to supply radon services, with uranium exploration largely replaced by home monitoring for indoor radioactivity.

RADON INDOORS

Measuring radon indoors is the most widespread use of track etching, and the one that the general public most often encounters–usually without knowing that track etching is involved or knowing how it works. Because the literature is vast, the reader is referred to books on the subject by Cohen (C87) and Nazaroff and Nero (NN88)–the first a popular work for the readers of *Consumer Reports*, the second a work with more technical detail.

As noted earlier, the ability of track etching to give a long-term average over the inevitable fluctuations of radon concentrations in a home makes it the method of choice (C87) for individual homeowners who have the patience to make measurements over a year or even a few months. A look at the variability of the statistics in different localities is helpful in giving guidance as to where homes should be tested most abundantly.

Figure 3-7, for example, shows results in Houston, Texas; Maine; and Sweden. Two features are noteworthy here. At each location there is a wide spread of values, and the most abundant ones are very different in the three cases. In this area of Sweden, where uranium-rich alum shale is the surface rock, most readings are rather high; among granitic, somewhat-high-uranium rocks in Maine, values are high–but less so; and over low-uranium, water saturated soils at Houston little radon penetrates indoors. Even with the observed spread of values, a small fraction of Houston homes are high, so that a reasonable tactic there is to test only homes with clear potential problems, such as bare-earth crawl spaces, basements with unsealed cracks, very tight above-ground structures, or badly cracked concrete slabs in contact with the ground. In contrast, high readings in the area tested in Sweden were so abundant that for safety all homes there should be tested. For more extensive data see Alter and Oswald (1987), who gave the first distribution of indoor readings by the track method for the United States.

Surveys with a few detectors in a single home can also be diagnostic, Figure 3-8 being an example. The value of 41 pCi/l in the basement, decreasing to 26 and 11 pCi/l on the ground floor and second story respectively, implies that radon was entering via the basement. And tests of the building blocks that made up the side walls against the soil subsequently showed them to be easy paths for radon, being both porous and uncoated.

It might also be noted that not everyone agrees that radon is a major, life-threatening hazard. There is a mine at Basin, Montana that offers to allow you to pay $2.50 per hour for the privilege of inhaling "health giving" radon.

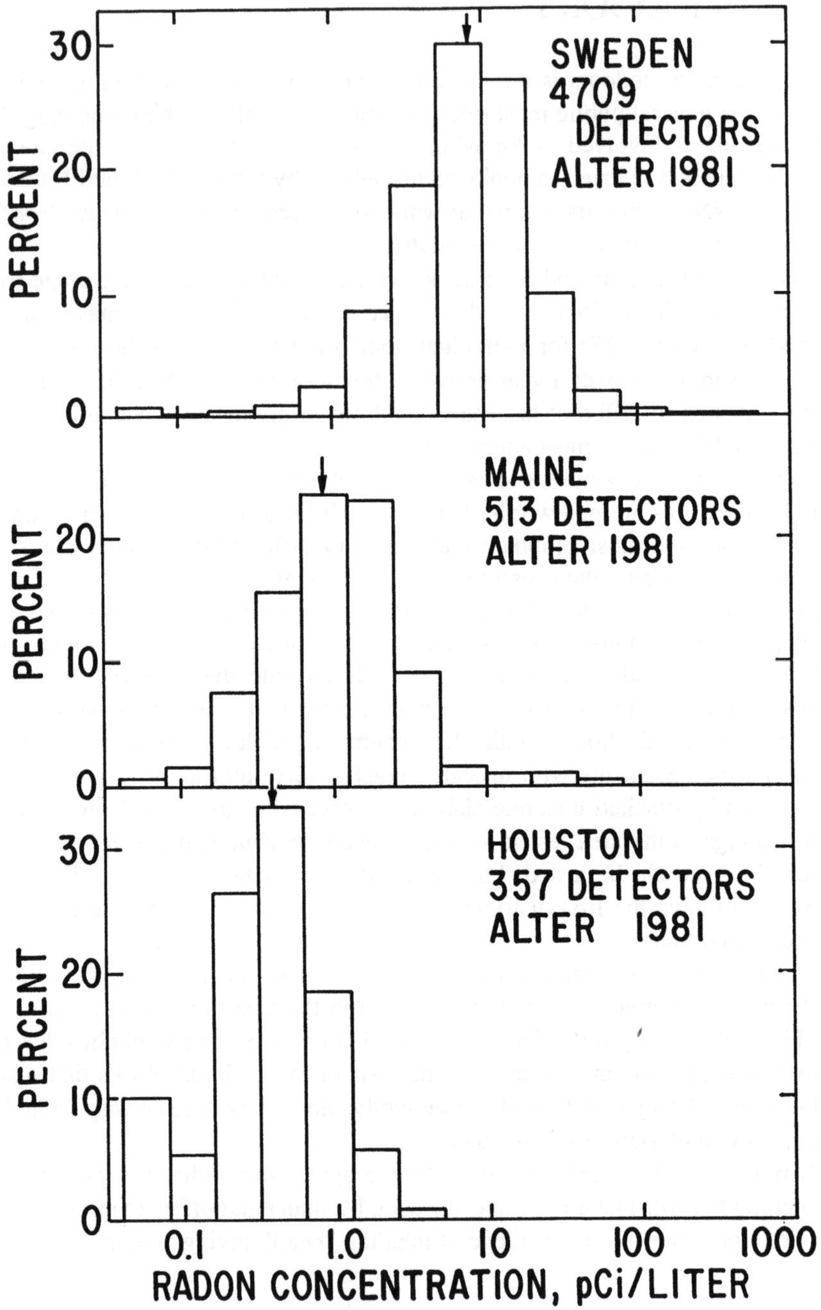

Figure 3-7. Comparison of indoor radon values in three locations. At each there is a dispersion, but average values shift upwards from Houston to Maine to Sweden because of increased uranium in the ground. It is estimated that residence for 60 years in a house with 10 pCi/l adds a 1.5% chance of dying of lung cancer.

	Feb-Apr 1980	Oct. 80-May 81	May-Oct 1980	Full Year
2nd Floor	11.0 ± 0.7	11.3 ± 0.7 12.1 ± 0.8	2.4 ± 0.2	7.1 ± 0.4
1st Floor	25.5 ± 1.6	12.2 ± 0.8	3.0 ± 0.2	7.6 ± 0.4
0.5 Floor		16.6 ± 1.1	4.5 ± 0.3	10.6 ± 0.6
Basement	40.6 ± 2.3	32.6 ± 2.1	9.1 ± 0.6	20.9 ± 1.1

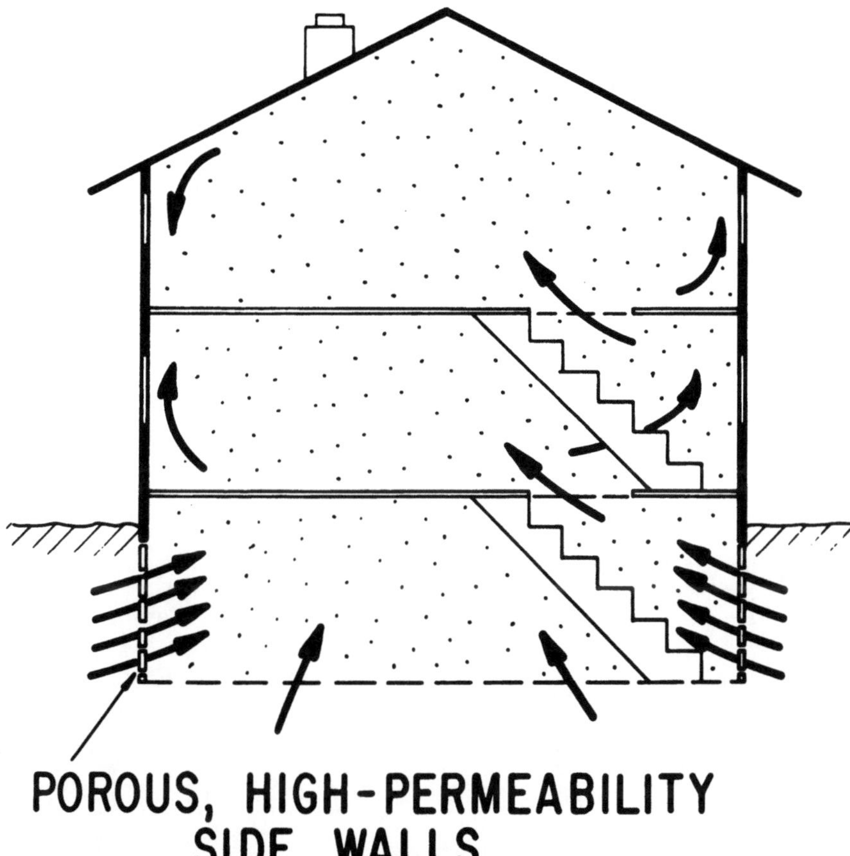

Figure 3-8. Tight home with porous basement side walls. Radon concentrations are given in pCi/l. Radon levels are higher in winter, when the house is more tightly closed. The downward increase in radon directs attention to the basement, where side walls admit excessive amounts of ^{222}Rn (F88).

On the other side of the ledger, the late Henry R. Hurwitz of General Electric insisted that the radon in his own home (1–2 pCi/l) was as threatening as "a nuclear meltdown" of a reactor. His point was that regulations restrict radiation to the public around reactors to levels that are far less than those readily accepted for radon in homes. Hurwitz is widely recognized for his insistence on containment for reactors in the U.S., which have served us well–as contrasted to the Chernobyl type of reactor that released much activity in Ukraine.

OIL AND GAS

The simplicity of the rural crossroads scene in Oklahoma in Figure 3-9 hides a wealth of radon variability in that region–a detailed set of earth messages about what lies below. Here the earth's surface hides a portion of the extensive Anadarko Basin oil and gas field at a depth of more than 7 km.

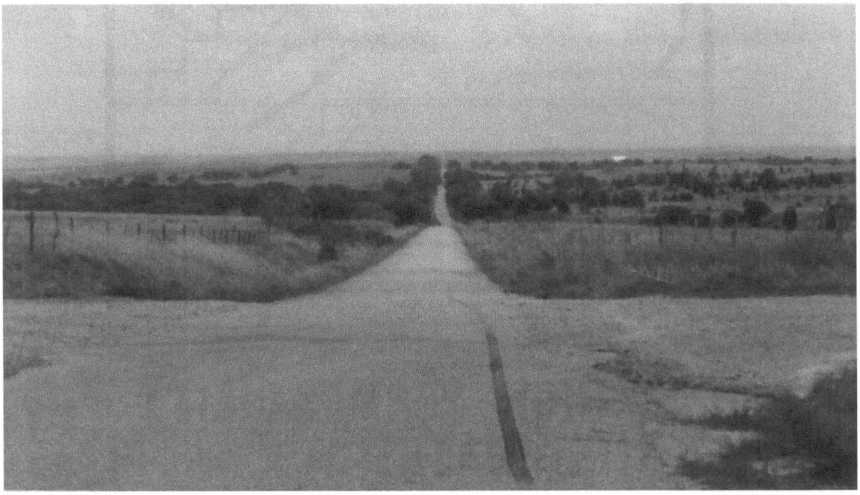

Figure 3-9. Oil country in Oklahoma is crossed by a grid of roads with 1-mile spacings. Crossroads such as this provided a convenient grid of sites for planting radon detectors.

At crossroads sites such as that shown in Figure 3-9, which locally comprise a regular grid with 1-mile spacing, radon monitoring at a 60-cm depth over two months gave the radon track densities shown in Figure 3-10 (Fleischer and Turner, 1984). The three highest readings lie over the high points given by the dashed lines for a second quantity – the distribution of oil wells and gas storage tanks that are used here as a measure of hydrocarbon resources that had been found over the years.

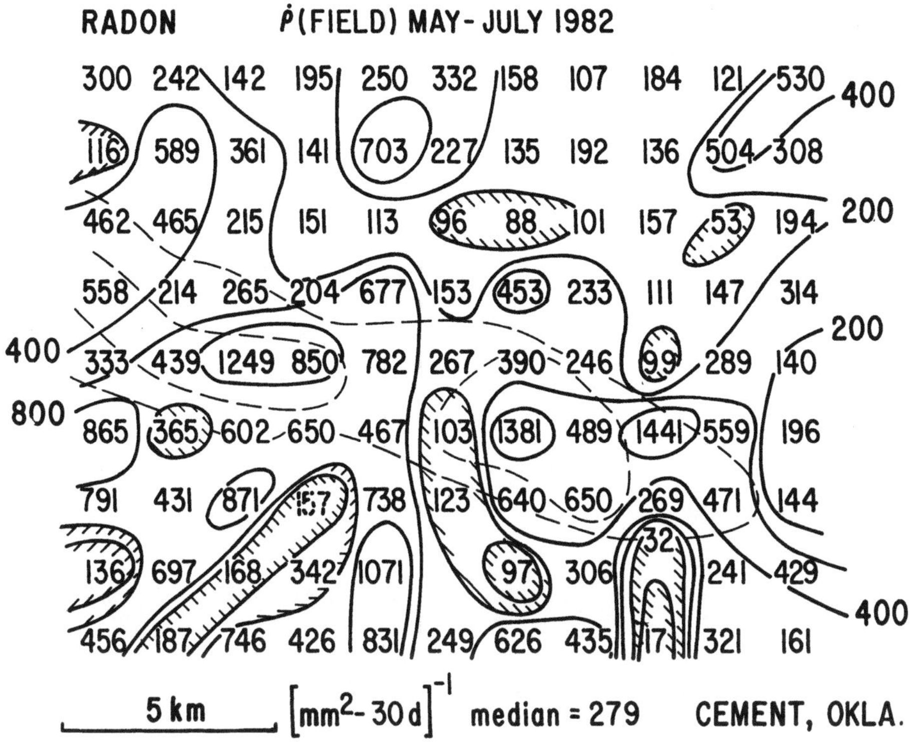

Figure 3-10. Comparison between location of petroleum and the ^{222}Rn values in the ground for a two-month period at Cement, Oklahoma. The numbers are values of the track-production rate. One track/(mm^2-30 days) equals 1.10 pCi/1 of ^{222}Rn. The three highest values lie at the peak of the superimposed density of oil and gas wells and storage tanks (WST) derived from USGS topographic maps and given here by the dashed lines. The contours are 25 and 50 WST/(miles)^2 values. (Fleischer and Turner, 1984).

The result in Figure 3-10 is satisfying but it is only a radon discovery of petroleum that had been discovered and developed by other methods long ago. I have a personal communication from H. W. Alter that a number of new finds have been made by this method. But oil companies tend to publicize neither their successes nor failures. As a result, the status is clouded by secrecy, and examples of such discoveries are not publicly available.

At any rate the results here give encouragement that soon we can be pumping more oil, Figure 3-11 being one strange example.

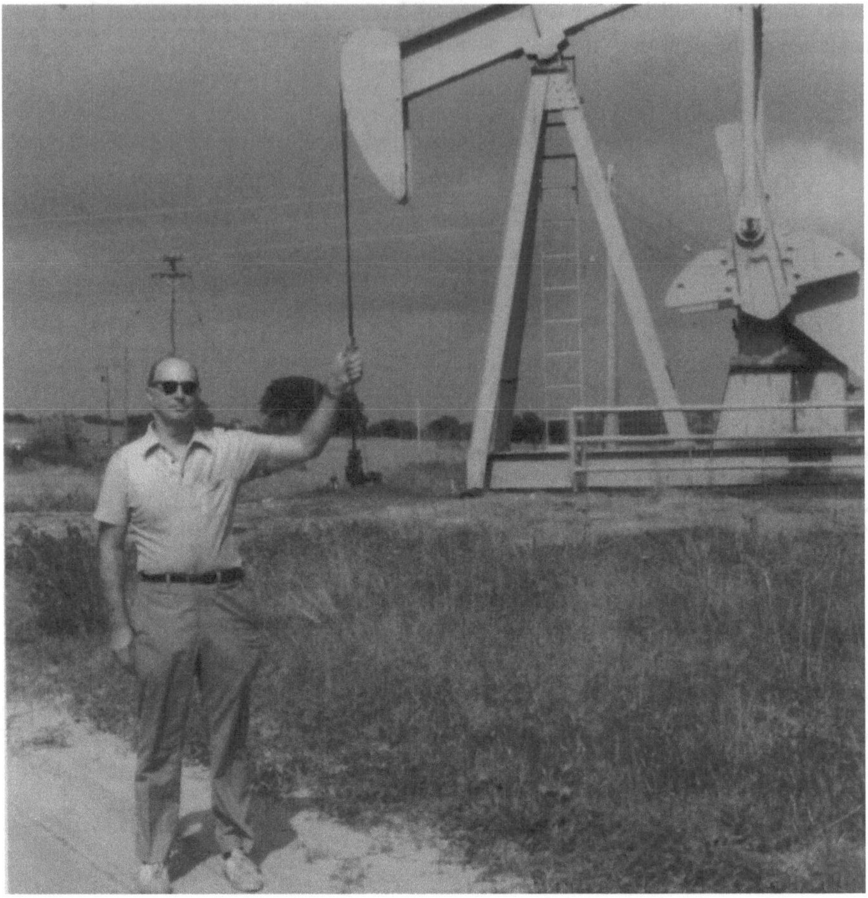

Figure 3-11. The author purportedly pumping oil. (The photograph with its skillful alignment is by L. G. Turner.)

Why might one expect radon concentrations to be elevated over hydrocarbon deposits? Two possibilities have been noted. The more direct, physical effect is that gas that leaks out of the deposit will move upward and thus create upflow at the surface where radon is measured, and this upflow distorts the near-surface radon values by the effect sketched in Figure 3-6. A second possibility is that there are chemical effects that occur where oil is produced naturally that may lead to uranium enhancements that are part of a geochemical halo around the oil and gas field. Field measurements of the soil indicate in this case that a minority fraction of the effect may be from such near-surface enrichment.

Fresh air and magnificent wildflowers in the Oklahoma spring make fieldwork of the sort done here into a welcome diversion from laboratory life. It also can stimulate adrenaline production. On one of our visits Turner and I had been interrupted in our work and were waiting in our vehicle for an intense rain shower to taper off. From turning on the car radio in an attempt to learn how long the precipitation was expected to continue, we learned that a tornado had just descended on a nearby town to the southwest and was last seen moving "toward the northeast." Carefully, but very rapidly, we deduced that it was time to retreat.

EARTHQUAKES

The calm of Blue Mountain Lake (Figure 3-12) belies the activity of multiple seisms that occurred in the 1970s in that region–the site of the first successful earthquake prediction in the United States (Scholz *et al.*, 1973). Separately, observations of radon entered the field of seismology around that time. Following the early report by Sadovsky *et al.* (1972), there have been numerous further observations in the ground of changes in radon that preceded earthquakes. But that report was what attracted Mogro-Campero's and my attention and encouraged us to visit Blue Mountain Lake to install radon detectors at this site (where seismologists from Lamont-Doherty Geological Observatory were monitoring conventional seismic information).

One of the Lamont people, Roger Bilham, accompanied us to point out where the seismic activity had been and show us where their instruments were located. As we were sogging through an open marshy area, we heard a sound like thunder, but oddly it was a clear day. Immediately a military jet passed overhead, suggesting to us that we had heard a sonic boom.

After emplacing our radon detectors, we entered the village of Blue Mountain Lake and encountered the postmistress, Mrs. Johnson, who knew of Bilham's interest in seismology. "Say, did you notice the earthquake?" she asked! Bilham phoned Lamont, 300 km away, to find that their Blue Mountain Lake seismometer did show a magnitude (M) 3.9 earthquake that had occurred 14 km from us. (In the earthquake magnitude scale each increase by one corresponds to about a 30-fold increase in the energy released.)

Monitoring for this event is incomplete, starting just after this significant earthquake, and it gave us a high reading that dropped down to what turned out to be the usual background level (Mogro-Campero and Fleischer, 1977).

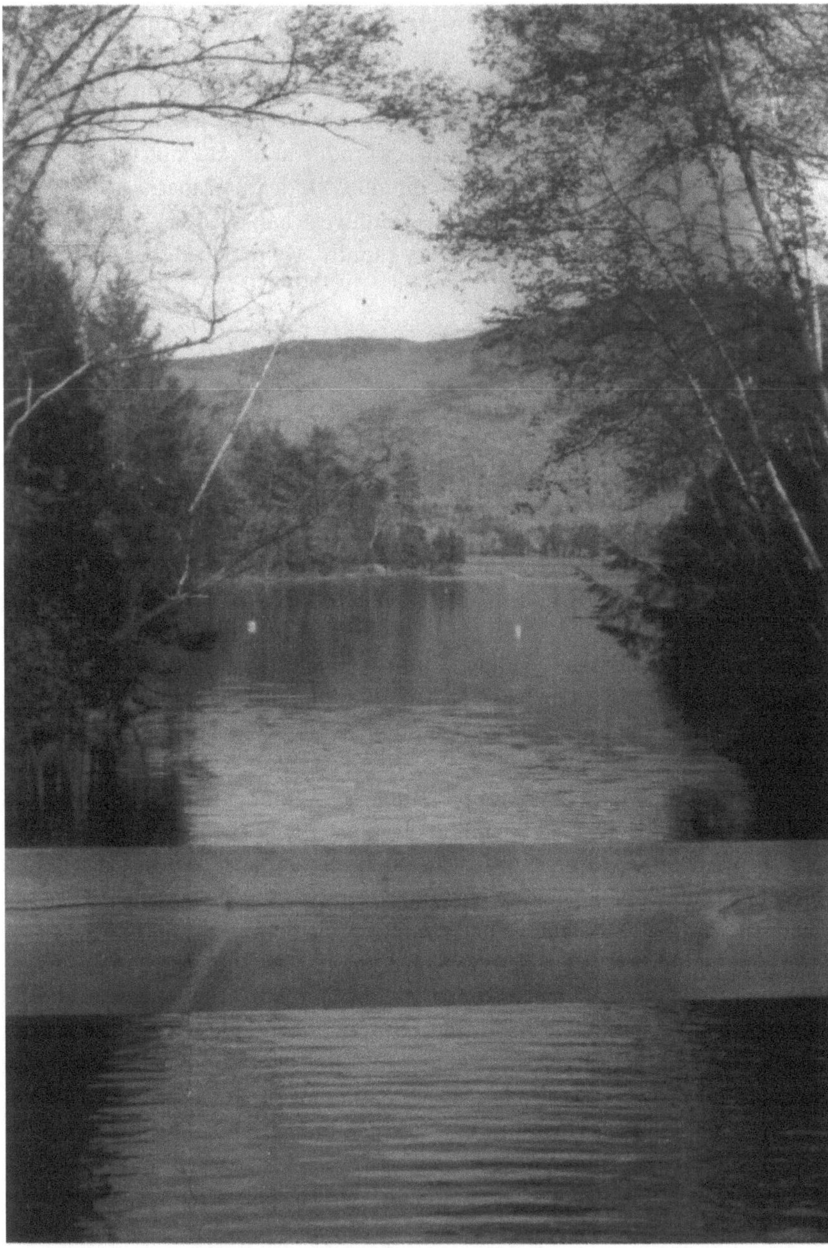

Figure 3-12. The serene site at Blue Mountain Lake, a site of seismic activity during the 1970s (photograph by L.G. Turner).

In contrast, an example of potentially useful radon observations is given in Figure 3-13. The figure contains four radon records surrounding the time of an M = 6.3 earthquake in Alaska. These stations were 178 km from the earthquake. Stations 1 and 3 showed radon increases of 40 times and 10 times, respectively, prior to the event, and then returned to their earlier levels. Premonitory changes at the other two stations are absent or unclear, but half of the sites are potentially useful.

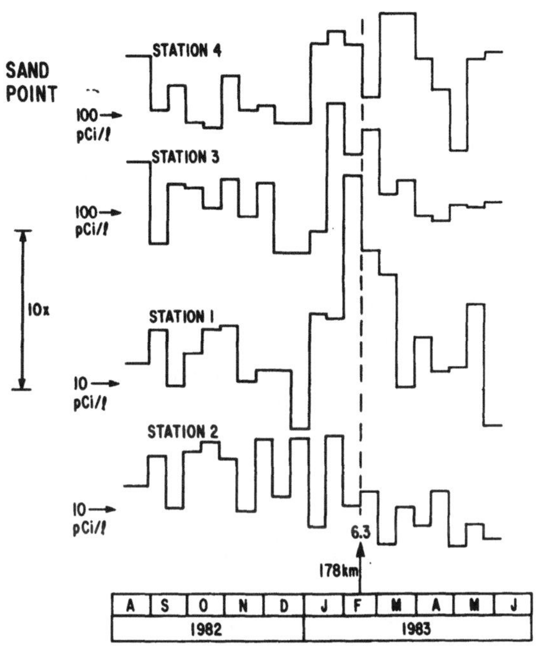

Figure 3-13. Radon readings at Sand Point, Alaska, from August 1982 to June 1983. Two of the four stations showed significant increases starting six weeks prior to the magnitude 6.3 event at a 178-km distance. The different magnitudes of the response for stations that are nearly equidistant from the earthquake (0, 6, 10, and 40 times increases) show the variability of station sensitivity and the need to assess each monitoring site (Fleischer and Mogro-Campero, 1985).

The effects at stations 1 and 3 have been interpreted as caused by the slight but long-range strains from the stresses that originate at the hypocenters of earthquakes in the making. Strains from a stress center in a solid will exist everywhere, but diminish strikingly with distance. Compressions or dilations will cause fluid flow through connected pore spaces and could thus alter near-surface radon–as sketched in Figure 3-6.

Two separate but related mechanical models have been proposed to explain radon effects (Dobrovolsky *et al.*, 1979; Fleischer, 1981); both give the same scaling relation as to how far from an event a given size earthquake might affect radon.

Figure 3-14 is an example of the expected scale, an example that suggests how such observations can best be used. Happily, the large events–the ones one most wants to predict–can influence radon most widely. Thus, multiple stations in a group would be expected to respond, the redundancy adding confidence that something truly is impending, and the center of the group identifying where the earthquake is likely to occur. A few stations can give overlapping coverage in a sizable area, as illustrated. The station locations are represented by the filled circles. The bounded area is the region in which an M = 6.2 ("moderate") earthquake should affect more than one station. Resolution and overlap expand rapidly for larger events.

A sad note for U.S. workers on radon and earthquakes is that the Geological Survey, which supported most of the research, has largely abandoned radon as a possible seismic predictor, because it is "not fully dependable." Those who follow earthquake prediction will realize that no method by itself has provided certainty or is likely to. Rather the long-range best hope for reliability is a combination of techniques, each adding to the probability of a correct inference. Thus, seismology could well adopt the attitude, mentioned earlier in this chapter, that has aided mineral exploration. We should use any information that serves to lessen uncertainty.

GAS FLOW IN THE EARTH

Flow of gas in the earth has been a unifying theme in the different areas of use of radon monitoring. For earthquakes, *changes* are what are sought, and they often correlate with stresses from seismic events. Oil and gas deposits may create their own flows by releasing gas. For uranium ores natural flows–possibly convective–can sometimes reveal deeper deposits than otherwise would be evident, and unidentified long-term flows over periods of many months have been observed (Fleischer and Mogro-Campero, 1979).

Another factor that deserves mention is the effect of alternating high- and low-pressure systems in the atmosphere, which Clements and Wilkening (1974) showed provide pumping that causes increases and decreases in the near-surface radon in the earth. For the uses in exploration and seismology, this effect provides unwanted noise–which is minimized by monitoring over times that are long enough to average over these alternations. For indoor radon the pumping can in some homes cause additional radon injection, and it thus increases potential health hazards.

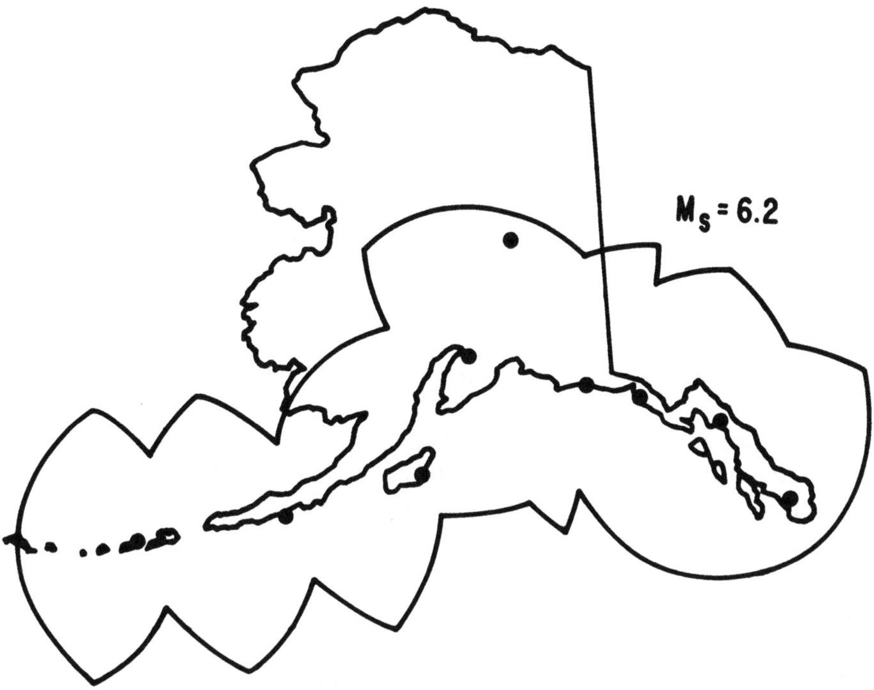

Figure 3-14. Zone of overlap in earthquake detection. Pre-seismic radon changes at more than one station are expected within the marked area for a 6.2-magnitude earthquake. Closed circles are the assumed locations of monitoring stations. Larger earthquakes would give rapidly increasing redundancy.

In the geological applications the increases in radon are best observed, as Figure 3-6 shows, by emplacing detectors at a depth of about 60 cm, where maximum changes occur from vertical flow. Although this consideration is the proper technical basis for choosing this depth, the actual original reason for this selection is different. Sixty centimeters is about the maximum depth from which the unaided average human arm can reach to retrieve a detector cup.

FURTHER READING ON RADON

General on radon and health:

C87 B. L. Cohen, *Radon, a Homeowner's Guide to Detection and Control*, Consumers Union, Mt. Vernon, NY, 1987, 215 pages.

NN88 W. W. Nazaroff and A. V. Nero, Jr., *Radon and Its Decay Products in Indoor Air*, J. Wiley and Sons, New York, 1988, 518 pages.

Radon and Tracks:

F88 "Radon in Environment–Opportunities and Hazards," R. L. Fleischer, *Nucl. Tracks and Radiation Measurements* **14**, 421–435 (1988).

For radon fundamentals, techniques, and applications:

ID97 *Radon Measurements by Etched Track Detectors*, R. Ilić and S. A. Durrani, ed., World Scientific, Singapore, 1997, 387 pages.

REFERENCES

H. W. Alter and R. A. Oswald (1987), "Nationwide Distribution of Indoor Radon Measurements—a Preliminary Data Base," *J. Air Pollut. Control Assoc.* **37**, 227–231.

H. W. Alter and P. B. Price (1968), "Radon Detection Using Track Registration Material," U.S. Patent 3,665,194, filed (1968), issued (1972).

W. E. Clements and M. H. Wilkening (1974), "Atmospheric Pressure Effects on ^{222}Rn Transport Across the Earth-Air Interface," *J. Geophys. Res.* **79**, 5025–5029.

I. P. Dobrovolsky, S. I. Zubkov and V. I. Miachkin (1979), "Estimation of the Size of Earthquake Preparation Zones," *Pure Appl. Geophys.* **117**, 1025–1044.

R. L. Fleischer (1981) "Dislocation Model for Radon Response to Distant Earthquakes," *Geophys. Res. Lett.* **8**, 477–480.

R. L. Fleischer, P. B. Price, R. M. Walker and E. L. Hubbard (1964), "Track Registration in Various Solid State Nuclear Track Detectors," *Phys. Rev.* **133A**, 1443–1449.

R. L. Fleischer and A. Mogro-Campero (1978), "Mapping of Integrated Radon Emanation for Detection of Long-Distance Migration of Gases within the Earth: Techniques and Principles," *J. Geophys. Res.* **83**, 3539–3549.

R. L. Fleischer and A. Mogro-Campero (1979), "Radon Enhancements in the Earth: Evidence for Intermittent Upflows?" *Geophys. Res. Lett.* **6**, 361–364.

R. L. Fleischer and A. Mogro-Campero (1985), "Association of Subsurface Radon Changes in Alaska and the Northereastern United States with Earthquakes," *Geochim. Cosmochim. Acta.* **49**, 1061–1071.

R. L. Fleischer and L. G. Turner (1984), "Correlations of Radon and Carbon Isotopic Measurements with Petroleum and Natural Gas at Cement, Oklahoma," *Geophysics* **49**, 810–817.

R. S. Likes, A. Mogro-Campero and R. L. Fleischer (1979), "Moisture-Insensitive Monitoring of Radon," *Nucl. Instr. Methods* **159**, 395–400.

A. Mogro-Campero and R. L. Fleischer (1977), "Subterrestrial Fluid Convection: a Hypothesis for Long-Distance Migration of Radon within the Earth," *Earth Planet. Sci. Lett.* **34**, 321–325.

M. Sadovsky, I. L. Nersesov, S. K. Nigmatulleav, L. A. Latynina, A. A. Lukk, A. N. Semenov, I. G. Simbireva and V. I. Ulomov (1972), "The Processes Preceding Strong Earthquakes in Some Regions of Middle Asia," *Tectonophysics* **14**, 295–307.

C. H. Scholz, L. R. Sykes and Y. P. Aggarwal (1973), "Earthquake Prediction: a Physical Basis," *Science* **181**, 803–810.

W. J. Ward, III, R. L. Fleischer and A. Mogro-Campero (1977), "Barrier Technique for Separate Measurement of Radon Isotopes," *Rev. Sci. Instrum.* **48**, 1440–1441.

B. Weidenbaum (1970), "Radon Enmometry Test with Track-Etch Film Rob Claims, Cities Service Minerals," General Electric, Vallecitos Atomic Laboratory, Report, October, 58 pages.

Chapter 4
TRACKS OF TIME

Time discovers the truth
Seneca, *De Ira, II*, c. 43.

Picture yourself to be an atom in a crystal. Life would be monotonous and claustrophobic, with constant jostling by your neighbors, just enough to keep you as warm as they are. Occasionally, something really exciting would occur if you happen to be near a uranium atom that spontaneously fissions, so that you and many of your neighbors are given sudden short-distance rides like those of the atoms sketched in Figure 1-8a. Between such events monotony takes over again.

Spontaneous fission was discovered in 1940 by Flerov and Petrjak in the nuclear species of interest here, ^{238}U. Spontaneous fission occurs in many of the synthesized trans-uranic elements, but is insignificant in any other of the natural elements found on earth.

Although the tracks from natural fission events form infrequently, they will accumulate, if they are stable, with the result that over geological times many may be present – their numbers increasing with the concentration of uranium *and the age of the solid.* As an example, if one part per million of uranium by weight is present in a mineral with the typical 6×10^{22} atoms/cm^3, on the average about 4 tracks/cm^3-year will form. Hence over typical geological times of millions of years, several millions of tracks/cm^3 will form. If all tracks are preserved, we need only to know the uranium concentration and the abundance of these natural tracks from spontaneous fission of uranium to determine the age of the mineral – the procedure being called fission-track dating.

This elegantly simple idea for track dating was spelled out by Price and Walker (1963), inspired by their 1962 observations of fossil tracks in mica that in nature had been immediately adjacent to high-uranium material. The first tests by the three of us with co-workers M. Maurette, D. S. Miller, P. Pellas, and E. M. Symes, which appeared in papers by Fleischer *et al.* (1964), Fleischer and Price (1964a), and Maurette *et al.* (1964), showed good agreement with established ages for some samples – but there were significantly low ages for a number of the samples that were used to test the method. At that time we worried that these discrepancies were a signal of some intractable defect in the method.

That fear proved to be extravagant. The "low" ages have turned out to be evidence of a special attribute of fission track dating – the ability to learn about ancient natural heating and cooling effects that include local igneous intrusions, regional mountain building, and baking of hydrocarbons. Each of these items is a matter of profound scientific, economic, or technological importance.

Incidentally, the able, versatile late geologist Henry Faul told me some years ago of his direct optical search for spontaneous fission tracks – with dating in mind. Unfortunately for him, his work was before intentional track etching was known and therefore his search failed. Interestingly, etched tracks had long ago been inadvertently observed by Baumhauer (1894) and others later, but the origin of the etched features was not known. Faul was the fun-loving editor of a valuable book on *Nuclear Geology*, the contributors to which he described accurately as "ranging from Phair to Faul." His student at Southern Methodist University, Charles W. Naeser, worked with Price, Walker, and me in 1964 and has gone on to be a major contributor in the field of fission-track dating.

THE METHOD

The procedure for dating can be nearly as simple and direct as was just stated. A mineral or natural glass is either cleaved or polished to expose a clean interior surface, and then etched to reveal natural tracks (as in Figure 4-1a). To measure the uranium content, the sample is irradiated by a specified dose of slow neutrons that induce fission of a portion of the ^{235}U atoms in the sample. Figure 4-1b shows the increased track density that resulted in the same natural glass as that in Figure 4-1a. The ratio of the original track density to that of the newly induced tracks gave in this case an age of 4.3 million years (M.y.) (Fleischer and Price, 1964b). It's that easy – in principle.

In practice there are numerous questions that have been resolved, in some cases differently by different workers. Do you look at the identical surface before and after you irradiate and re-etch? If so, it must be kept meticulously clean from uranium contamination. If you don't use the same surface, you must show that the uranium concentration doesn't vary from place to place in your sample. Do you calculate the induced track density as the difference between the total after irradiation and the original density of fossil tracks? Or, to avoid having increased error from a difference, do you erase the natural tracks by heating the sample prior to irradiation – and in so doing risk altering its etching properties? How do you measure the neutron dose simply and accurately? We will answer this last question in Chapter 6. Are there significant effects of past exposure to heat, pressure, and deformation? Are there etched features that aren't tracks but might be mistaken for tracks? Are there other tracks besides those we wish to use for dating? Are tracks permanent, once formed?

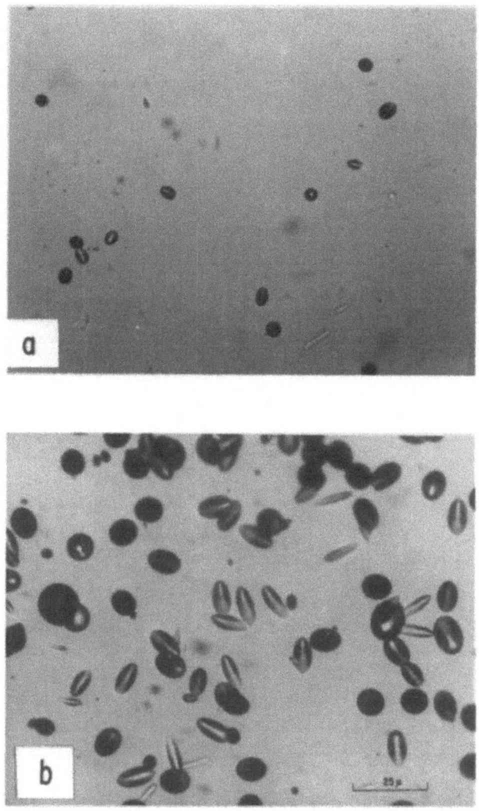

Figure 4-1. Fission track dating. The finding of natural tracks from the spontaneous fission of trace amounts of ^{238}U in materials such as the obsidian (a) has allowed the ages of minerals to be measured. The uranium content is measured by inducing further fission by a neutron irradiation as in part b. The obsidian was etched in hydrofluoric acid.

All these questions have been faced and answered. For many of the details the reader is referred to more complete sources such as FH72, FPW75, WH92, and to issues of the journal *Nuclear Tracks* that contain papers from the regular fission-track dating conferences that were held in Pisa in 1980, Troy (1984), Besancon (1988), Philadelphia (1992), and Ghent (1996). As an indication that other features superficially resemble tracks, but can be easily rejected, Figure 4-2 shows a glass with bubbles and a single fission track that is black because it does not scatter light back into the microscope due to its conical shape. Not all centers are dark, as seen in Figure 4-1, but those with light centers have distinctive shapes.

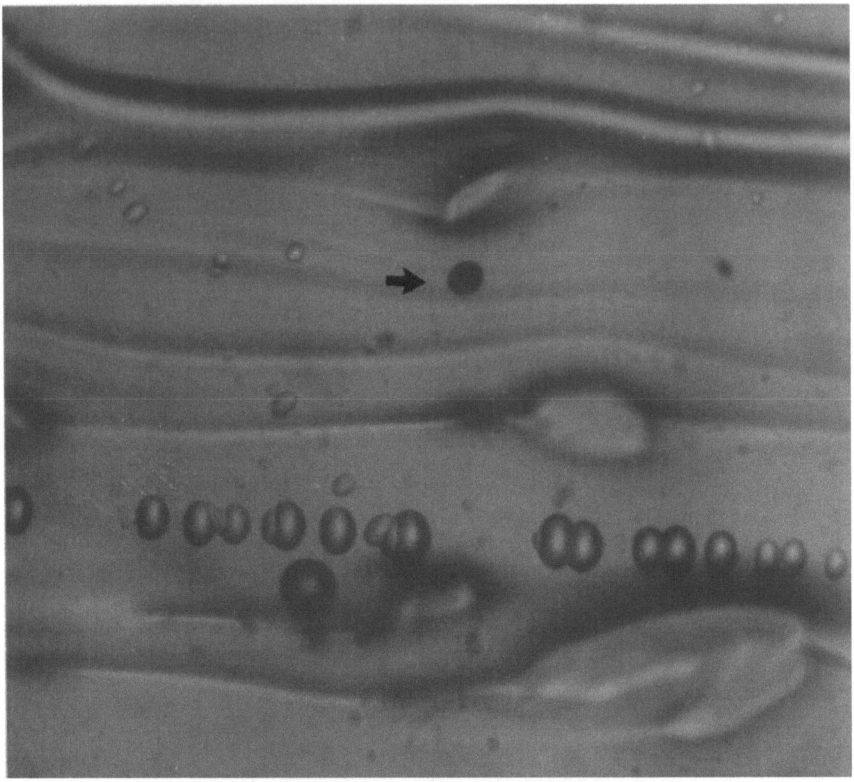

Figure 4-2. Example of background defects in natural glass. The round features with light centers are pores that reflect from the flat surface at the bottom. An arrow indicates the one etched track, which is distinctive and thus easily recognized.

As to other sources of natural tracks, suffice it to say that for samples that have always been on earth, the sole significant source of natural tracks is spontaneous fission of ^{238}U, except in very remarkable circumstances for ancient samples. The one documented exception is where induced fission was found to be present in nature as a result of the earliest nuclear reactors – at the site that is now the Okla mine in Gabon. These were natural reactors that operated without the intervention of man, in fact long before man had evolved. They were possible because 2,000 M.y. ago uranium was enriched to 3% ^{235}U, the amount that is normally used in manmade power reactors. Why was uranium enriched? The reason is that ^{235}U has a natural decay rate that is six times larger than that of ^{238}U and of the right magnitude that going back in time 2,000 M.y. its abundance was exponentially greater by a factor of more than four. Maurette (1976) describes the Okla phenomena from the point of view of those who are interested in nuclear tracks.

For samples that have spent time in space the situation is enlivened, but complicated, by many other varieties of tracks becoming possible – mostly from cosmic rays or their interactions in minerals, some of which will be briefly mentioned in the next chapter.

EARLY RESULTS

While we were testing fission-track dating, the major question was whether correct ages are found, "correct" being defined for geological samples by ages determined by other methods. For example, in potassium-argon dating, the amount of ^{40}Ar that accumulates in a mineral as a result of ^{40}K decay is measured. Different dating methods are most easily applicable to samples of different compositions; and they give checks on one another, since each has its own assumptions. In addition, fission-track dating is of special importance in measuring effects of ancient heating. Figure 4-3 is basically a score sheet for the early track dating. Points on the diagonal line show agreement between track dating and other methods, which for some of the younger samples are actual recorded dates of laboratory or commercial preparations of glasses. The overall impression is that the method works since there is good agreement for most samples; yet there exist determinations where track ages systematically are younger than the reference ages. On this log-log plot the discrepancies may not look large; but some ages are low by as much as a factor of three, i.e., far beyond statistical uncertainties. These are real discrepancies that we needed to understand. Now I will give some examples of the earliest opportunities we had to provide new results. Some of the notes are taken from the author's article in the informal fission-track-dating newsletter *On Track* (1993).

Corning Museum of Glass

The youngest group of glass samples in Figure 4-3 were ones we obtained at the Corning Museum of Glass with the collaboration of Robert H. Brill, the science curator (Brill *et al.*, 1964). For very young samples (with ages as low as 20 to 100 years) it is critical for the sample to be unusually high in uranium in order to have enough tracks to count. Fortunately there were periods when high-U glasses were created, first to produce a pretty yellow-green tint in decorative glass (the color being unfortunately absent in the photo in Figure 4-4) and second, commercially, to adjust thermal expansion coefficients to allow reliable

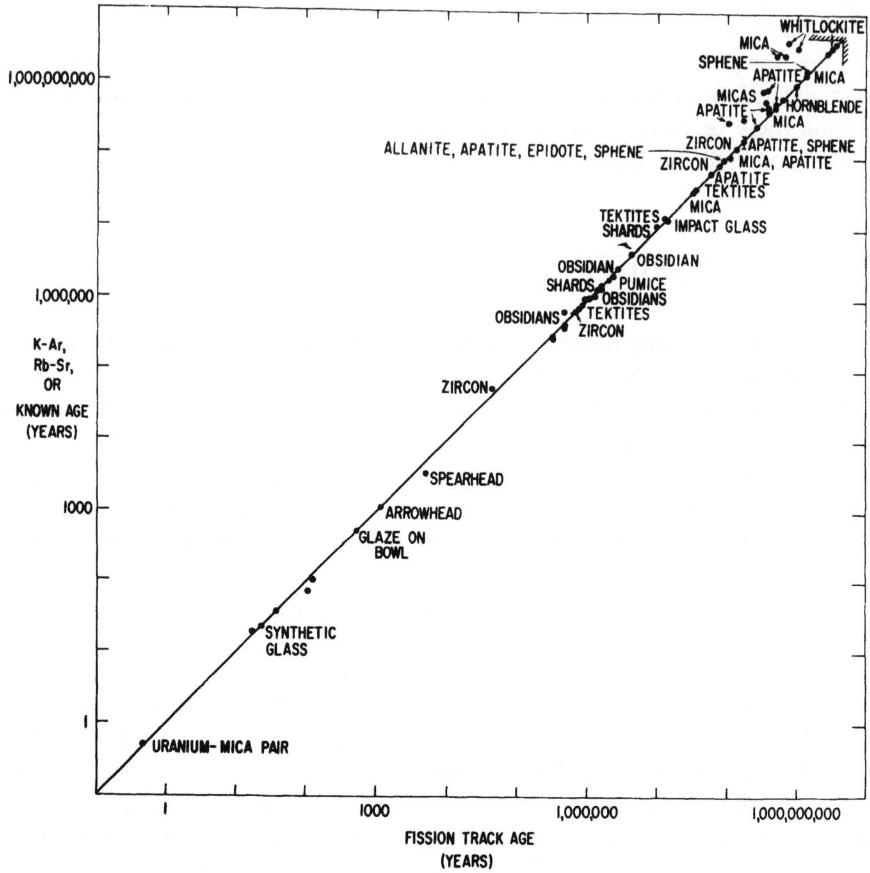

Figure 4-3. Fission track dating has been shown to give correct ages over time spans ranging from half a year to more than a billion years. The graph compares fission track ages, measured by counting natural tracks formed by spontaneous fission of ^{238}U, with ages known by other means, either documented ages for manmade samples that appear in the lower left or ages measured by other radioactive decay techniques for the geological samples indicated in the upper right. Occasionally fission track ages of natural samples are lowered by thermal effects, allowing the fission track technique to serve as a thermometer for past events. (FPW75)

glass-to-metal seals. Our technique was to survey likely-looking glasses in the museum with a radiation meter that I brought along from General Electric. While we were so engaged, we received a number of puzzled, wary glances from members of the public who where there to view the exhibits, but were not seeking radioactivity.

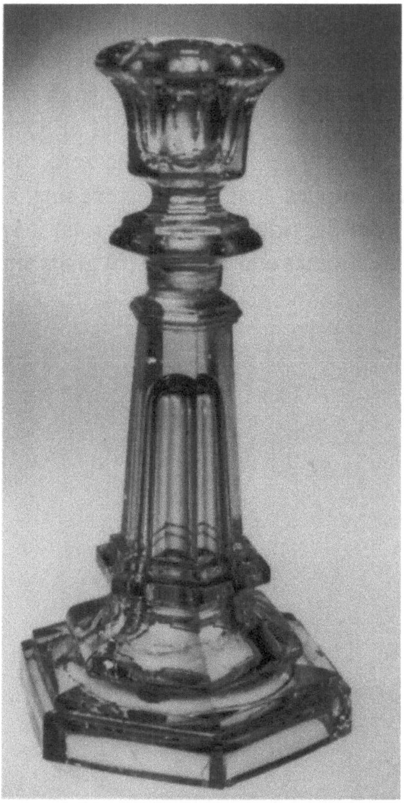

Figure 4-4. American candlestick by the Boston and Sandwich Glass Co., circa 1850–1860, containing 0.61 wt % UO_3. Fission-track dated at 1840 ± 20 (Brill *et al.*, 1964), i.e., 120 years old when dated.

Sea Floor Spreading

One of the great modern realizations in geophysics is that new ocean floor has been continuously created at midoceanic ridges and that the continents have been moving apart as evidence of major convective motion in the mantle of the earth. Although the discovery and the most extensive measurements of the nature and extent of ocean bottom spreading were inferred primarily from magnetic records in ocean bottom rocks of the polarity of the earth's magnetic field, the most direct measure of the rate of spreading has come from a combination of fission-track and K-Ar dating (Fleischer *et al.*, 1968a). Where spreading originates, molten basalt erupts on the sea floor, and a glassy rind is produced when the hot lava is quenched by the cold deep-ocean water, as is seen in Figure 4-5 – taken one mile beneath the ocean surface. The 4°C ocean bottom temperature preserves subsequent tracks and retains the argon from potassium decay. A very young fission-track age, 20,000 years, was found at 45°N in the Mid-Atlantic close to the spreading center – geologically time zero. And a value of 300,000 years some 7 km west implies that recent spreading has been at a rate of 2.5 cm/yr to the west from the ridge, or 5 cm/yr total if spreading is symmetrical. More recent work by Storzer and Sélo (1976) gives similar results at 37°N.

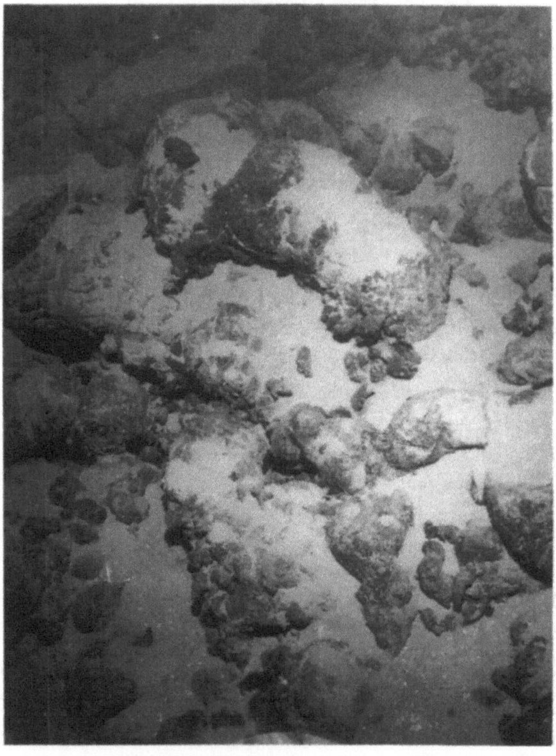

Figure 4-5. Pillow lava extruded at the floor of the Mid-Atlantic at 45°N latitude 5,000 feet below the ocean surface about 10,000 years ago. The glassy surface was dated by fission tracks (courtesy of F. Aumento).

In 1968, Trans World Airlines was advertising that they had made Europe closer to the United States than ever. The geochronological evidence, however, establishes a diametrically opposite effect.

Early Man

Olduvai Gorge, Tanzania, is a famous archaeological site where remains of two hominid species *Zinjanthropus* and *Homo habilis* were found in a stratum known as Bed I. A K-Ar age of 1.75 million years (Leakey *et al.*, 1961) was questioned, not because of the technical quality of the determination, but because anthropologists doubted that man could have been so advanced as the remains indicated for that early date. It was therefore clear that an independent dating method could be of use in resolving doubts.

The volcanic pumice of Bed I included glass shards that retained tracks, but the tracks were in low abundance and needed to be carefully distinguished from effects of etched porosity. Figure 4-6 shows that they can be recognized by one who is practiced, but the ability to view the sample alternatively in transmitted and reflected light is critical to avoiding counting spurious features. It is obvious that tracks in the pumice that was used were hardly obvious even to those familiar with tracks in glass, and major corrections were necessary to allow for empty space where tracks could not be observed. The result we arrived at, 2.03 ± 0.28 M.y. (Fleischer *et al.* 1965a), in spite of its sizable uncertainty, confirmed the K-Ar result, which is no longer questioned.

With respect to publication, mostly we had found that the new results of early track work breezed through journal review systems. One exception was this work with Louis Leakey on dating at Olduvai Gorge. At Leakey's urging we submitted the article to his favorite journal, *Nature*. We thought that the article might encounter questions about track identification and track-density corrections, but what I suspect we brushed into was a reaction from an archaeologist who intuitively disagreed with the (to him) great K-Ar age of 1.75 M.y. and our confirmation of it. I remind the reader that Price, Walker, and I were at that time part of a wide-ranging unit at General Electric that went by the not very descriptive name Metallurgy and Ceramics Department. Without explanation the editor returned the manuscript and suggested that we submit the article to a metallurgy journal. We decided that, as far as that editor knew, *Science*, where the article finally was published, was such a journal.

Tektites—Subjects of Dispute

Tektites are glass objects that have rained down on the earth in at least five discrete showers at various times in the past, usually on well defined areas, such as Texas, Czechoslovakia, or SE Asia plus Australia. Their origin is argued but their groupings helped us to make tests of track dating, which mostly agreed with K-Ar ages. For some Australian tektites the ages disagreed with ages inferred from the geological strata in which they are found. Figure 4-7 gives a sampling of their distinctive shapes.

These objects have been the subject of much wrangling. This situation became apparent to me at the first scientific meeting at which I presented dating results, the Second International Symposium on Tektites at Pittsburgh in 1963. Here, people with chemistry-related backgrounds (mostly) favored terrestrial origins, and workers with physics or aerodynamics training (mostly) leaned toward the moon. The bizarre and memorable low point of this affair was a poll at the end of the meeting in which the moon defeated the earth 21 to 18. That margin would not be convincing at a football game; it expresses the degree of unanimity that existed. This subject remains one of unresolved, polarized disputes.

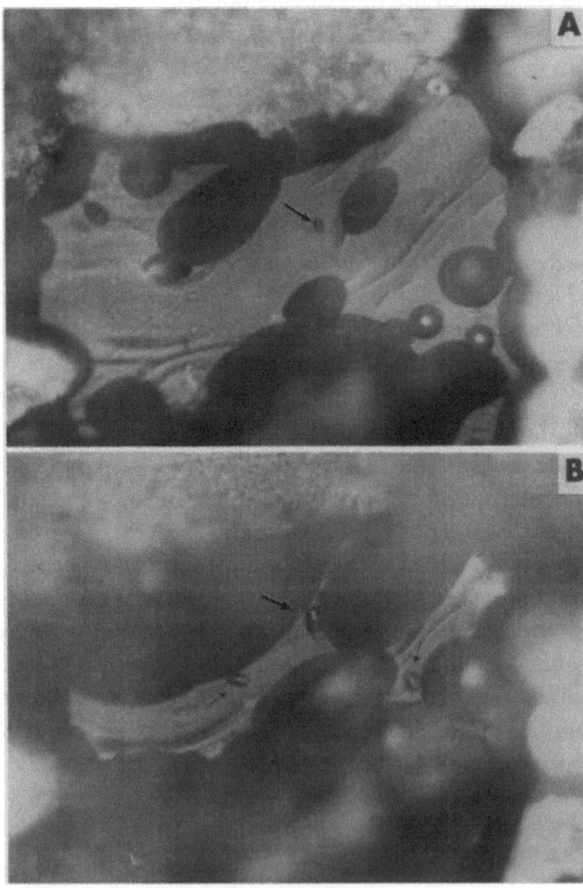

Figure 4-6. Spontaneous and induced tracks in pumice from Olduvai Gorge. The view in A shows a single spontaneous fission track (solid arrow); the view in B shows the same area after neutron irradiation and a 500-second immersion in 5% HF. The pit from the spontaneous fission track is now flat-bottomed, and two new tracks (thin arrows) have been induced by the irradiation (Fleischer *et al.*, 1965a).

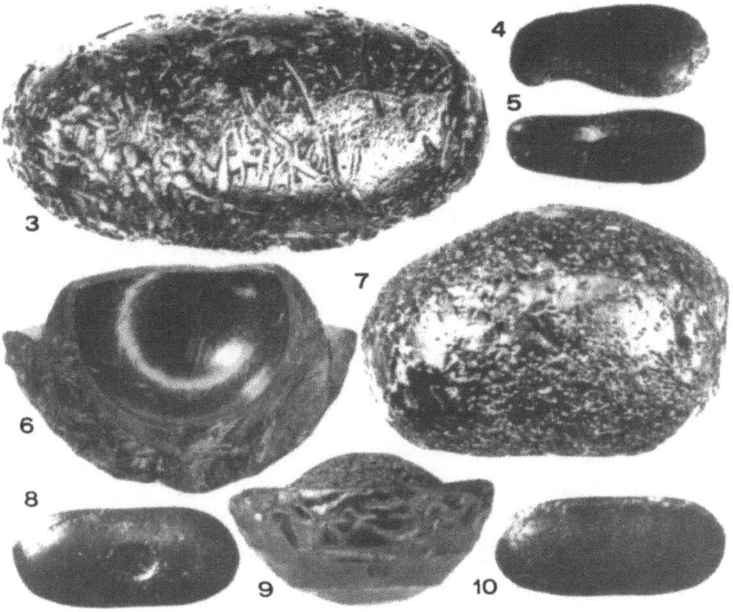

Figure 4-7. Tektites have diverse shapes, the most informative of which are similar to number 9. In entering the earth's atmosphere it adopted a stable position in which the front surface was ablated, with melted glass flowing around the pre-atmospheric spherical part, which is on the top as shown (O'Keefe, 1963).

THERMAL EFFECTS AND THERMAL DATING

Early in the tektite work we realized that heat affects tracks in a thermally activated way – short times being required to erase tracks at high temperature and much longer times at lower temperature (see Figure 4-8). The relation measured in the laboratory is commonly extrapolated far beyond experimentally accessible times to estimate whether tracks are stable over geological times. At the floor of the deep ocean the water temperature is low and remarkably stable, and therefore the inference from Figure 4-8 is that in this particular site tracks in the glass studied would be preserved over all of time.

Individual minerals or glasses have their own thermal fading characteristics; and thus for each, data similar to those in Figure 4-8 need to be measured to help validate age studies. Given the large number of minerals in which tracks can be etched (Table 4-1), proper documentation involves significant effort. Fortunately most track dating is done on a few well-documented minerals, for example, apatite, zircon, sphene, allanite, epidote, and various glasses. Many natural substances, however, have wide ranges of possible composition with very different track-annealing behavior. Apatites and glasses are examples of such materials – where annealing experiments on multiple compositions need to be done in order to understand thermal effects reliably (Crowley *et al.*, 1990, Green *et al.*, 1989).

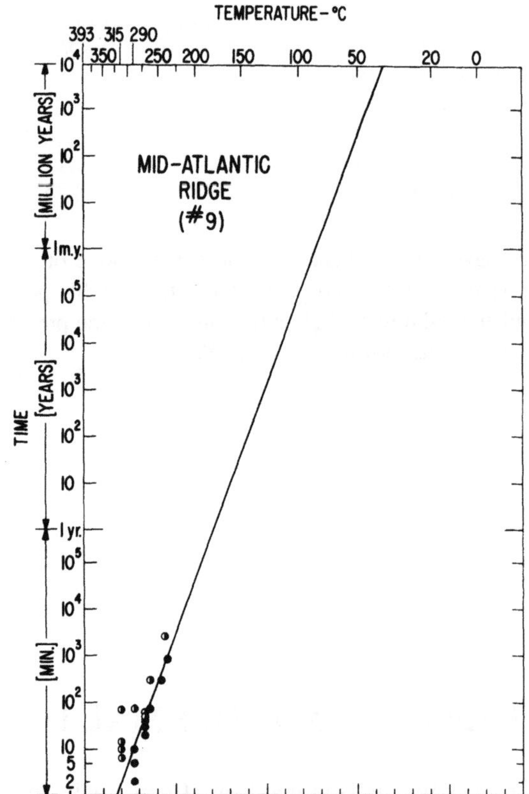

Figure 4-8. Track-fading conditions for a Mid-Atlantic ocean-bottom basaltic glass sample from 7.6 km west of the Median Valley. Solid dots indicate time and temperature conditions for which tracks are unaffected. Open dots indicate conditions when tracks are reduced in number or are absent (Fleischer *et al.* 1971).

Table 4-1. Minerals in Which Fission Tracks Can Be Etched

- Allanite $[H_2O \cdot 4(Ca,Fe)0 \cdot 3(Al,Fe)_2O_3 \cdot 6SiO_2]$
- Amber@
- Anthophyllite $(Mg,Fe)SiO_3]$**
- Apatite $[Ca_5(F,Cl)(PO_4)_3]$**
- Aragonite $(CaCO_3)$
- Autunite $[Ca(UO_2)_2P_2O_8 \cdot 8H_2O]$
- Barite $(BaSO_4)$
- Barysilite $(Pb_3Si_2O_7)$*
- Bastnäsite $(CeFCO_3)$**
- Benitoite $(CaTiSi_3O_9)$
- Beryl $(Be_3Al_2Si_6O_{18})$*
- Bismutite $(Bi_2O_3 \cdot CO_2 \cdot H_2O)$
- Brewsterite $[(Sr,Ba,Ca)O \cdot Al_2O_3 \cdot 6SiO \cdot 5H_2O]$
- Calcite $(CaCO_3)$*@
- Celsian $(BaAl_2Si_2O_8)$
- Cerussite $(PbCO_3)$
- Chabazite $(CaAl_2Si_4O_{12} \cdot 6H_2O)$**
- Chlorite $[(Mg,Fe)_5(Al,Fe)_2Si_3O_{10}(OH)_8]$**
- Clinopyroxenes $[Ca(Mg,Fe,Al)(Al,Si)_2O_6]$
- Cryolite (Na_3AlF_6)
- Epidote $[Ca_2(Al,Fe)_3(SiO_4)_3OH]$**
- Eulytite $[Bi_4(SiO_4)_3]$
- Feldspars $[(Na,K)AlSi_3O_8]$ $[CaAl_2Si_2O_8]$*
- Garnets $(Mg, Ca)_3(Fe,Al)(SiO_4)_3$**
- Gillespite $(FeO \cdot BaO \cdot 4SiO_2)$
- Glasses (many)**
- Glauconite $[K(Fe,Mg,Al)_2(Si_4O_{10})(OH)_2]$
- Gypsum $(CaSO_4 \cdot 2H_2O)$
- Halite (NaCl)
- Harmotome $[Ba_5(NaK)Al_{11}Si_{29}O_{80} \cdot 25H_2O]$
- Hardystonite $(Ca_{1.99}Pb_{.01}\ ZnSi_2O_7)$
- Heulandite [a zeolite, $(Ca,Na_2)O \cdot Al_2O_3 \cdot 9SiO^2 \cdot 6H_2O$]**
- Hornblende $[NaCa_2((Fe,Mg,Al)_5(Si,Al)_8O_{22}(OH)_2]$**
- Huebnerite $(MnWO_4)$
- Kleinite $[Hg_2N(Cl,SO_4) \cdot nH_2O]$
- Kyanite (Al_2SiO_5)**
- Leuchtenbergite $(H_8Mg_5Al_2Si_3O_{18})$
- Lithium Fluoride (LiF)
- Margarite $[CaAl_4Si_2O_{10}(OH)_2]$
- Melilite $[Ca_2(Al_2,MgSi)SiO_7]$*
- Micas
  - [Biotite, $K(Mg,Fe)_3AlSi_3O_{10}(OH)_2$]**
  - [Lepidolite;Zinnwaldite,$K_2Li_3Al_4Si_7O_{21}(OH,F)_3$]**
  - [Muscovite, $KAl_3Si_3O_{10}(OH)_2$]**
  - [Phlogopite;Lepidomelane, $KMg_2Al_2Si_3O_{10}(OH)_2$]**
- Melilite $[Ca_2(Al_2,MgSi)SiO_7]$*
- Microlite $(Ca_2Ta_2O_7)$
- Mimetite $[Pb_5Cl(AsO_4)_3]$
- Monazite $[(Ce,La,Th)(PO_4,SiO_4)]$*
- Mullite $(Al_6Si_2O_{13})$
- Nasonite $[Pb_4(PbOH)_2Ca_4(Si_2O_7)_3]$
- Nickel Halides $(NiCl_2, NiBr_2)$
- Olivine $[(Mg,Fe)_2SiO_4]$
- Orpiment (As_2S_3)*
- Orthopyroxene $(Mg_{1-f}Fe_fSiO_3)$
- Pennine (a chlorite)
- Pollucite $(H_2O \cdot 2Cs_2O \cdot 2Al_2O_3 \cdot 9SiO_2)$
- Prehnite $(Ca_2Al_2(OH)_2Si_3O_{10})$
- Pucherite $(BiVO_4)$
- Pyromorphite $(Pb_5Cl(PO_4)_3)$
- Quartz (SiO_2)
- Raspite $(PbWO_4)$
- Sanbornite $(BaSi_2O_5)$
- Scheelite $(CaWO_4)$*
- Sphene $(CaTiSiO_5)$**
- Spinel $(MgAl_2O_4)$
- Spondumene $(LiAlSi_2O_6)$
- Stilbite [$(Ca,Na_2)O \cdot Al_2O_3 \cdot 6SiO_2 \cdot 6H_2O$, a zeolite]**
- Stibiotantalite $[(SbO)_2(Ta,Nb)_2O_6]$
- Talc $[Mg_3Si_4O_{10}(OH)_2]$
- Thorite $(ThSiO_4)$
- Torbernite $[Cu(UO_2)_2P_2O_8 \cdot 12H_2O]$
- Tridymite (SiO_2)
- Topaz $[Al_2SiO_4(F,OH)_2]$
- Tourmaline (complex silicate)
- Vanadinite $[Pb_5Cl(VO_4)_3]$
- Vermiculite (biotite-derived**
- Vesuvianite $(Ca_6[Al(OH,F)]Al_2(SiO_4)_5)$**
- Whitlockite $\{Ca_3(PO_4)_2]$**
- Willemite (Zn_2SiO_4)
- Zoisite $[Ca_2(AlOH)Al_2(SiO_4)_3]$**
- Zircon $(ZrSiO_4)$**

*Natural tracks found.

**Fission track dating reported (mostly WH92 or FPW75).

@Track fading is a problem.

In Table 4-1 minerals are marked to indicate whether they actually have been used for dating (**), or whether natural fission tracks have been observed (*) – which would make them candidates for dating. Those where track fading is cited as a problem (@) are ones in which tracks fade at common environmental temperatures. In those with no footnotes, natural, internally produced tracks have not been seen for various reasons. Either they are low in uranium (quartz), they are manmade samples (LiF), or tracks may have faded, but fading conditions have not been measured.

Recognizing Thermal Effects

Rarely can one be confident that a material has spent its life at a known, well-controlled temperature, such as can occur for ocean bottom material. On the surface of the earth, solar heating is present, and natural fires can spread; beneath the surface the earth becomes progressively warmer, and local or regional volcanic activity can provide additional heat pulses or sustained heating.

When assorted ages are found for a group of samples that should be identical, but are not, heating of some of the samples is suspected. For example, the Australian tektite ages for different samples in Table 4-2 are expected to be about 700,000 years, but 4 of 9 are distinctly lower and different from one another (Fleischer and Price, 1964b). The inference is that those samples were heat-affected and that the group with higher ages defines the correct original age of about 700,000 years.

Table 4-2. Fission-Track Ages of Australian Tektites

Location	Age (yr)
Kalgoorlie	<30,000
Port Campbell, Pulpit Rock	110,000
Port Campbell	130,000
Kalgoorlie	340,000
Charlotte Waters	640,000
Kalgoorlie	660,000
Port Campbell	690,000
Port Campbell	690,000
Port Campbell	800,000

Figure 4-9 shows a clear case, and one of anthropological interest. Mesolithic man cleaved a knife from obsidian – presumably an old obsidian – and later it was placed in a hearth that was hot enough that it sagged under its own weight. Since such heating will remove previous tracks, those that are present now have formed since the fire died down. Its age of about 4,000 years (Fleischer *et al.*, 1965b) is consistent with the culturally inferred age of the site.

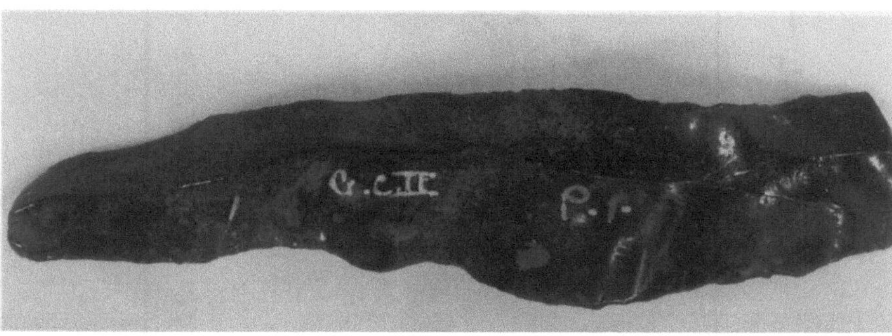

Figure 4-9. Burnt obsidian blade, 7.3 cm long, from Gamble's Cave II, Elmenteita, Kenya. The glass was resoftened by heating about 4,000 years ago as determined by counting the tracks formed since that heating removed the preexisting tracks (after Fleischer *et al.*, 1965b).

A more widely applicable method of recognizing heat-affected tracks was discovered by Storzer and Wagner (1969). Moderate heating does not totally erase tracks; it lowers the intensity of damage and thus the etching rate. The result is that the standard etching time for normal tracks produces fewer, shorter, blunter, smaller-diameter tracks (as was illustrated by Figure 1-3). The effect becomes obvious if one compares the sizes of the natural tracks with those of the fresh ones that are produced by irradiation with neutrons. As the diameters of heated-and-then-etched tracks in glass decrease, so too do the track densities, and they diminish in a well-defined way that is shown in Figure 4-10. Similar behavior to that in glasses is observed in crystalline minerals—where annealing effects appear as shortening of tracks (as will shortly be described).

Age Corrections

The behavior described by Figure 4-10 makes it possible to correct heat-reduced ages when tracks are not totally erased. From the track-diameter reduction the amount of the track-density reduction is found (by using track-annealing data like those in Figure 4-10) and can be used to infer the original track density. Presumably, if this procedure had been known in 1964, three of the four low ages in Table 4-2 could have been revised to give the proper value; the <30,000-year age, however, was a case where no tracks were found, and hence its age is not correctable. That sample presumably was heated strongly and recently.

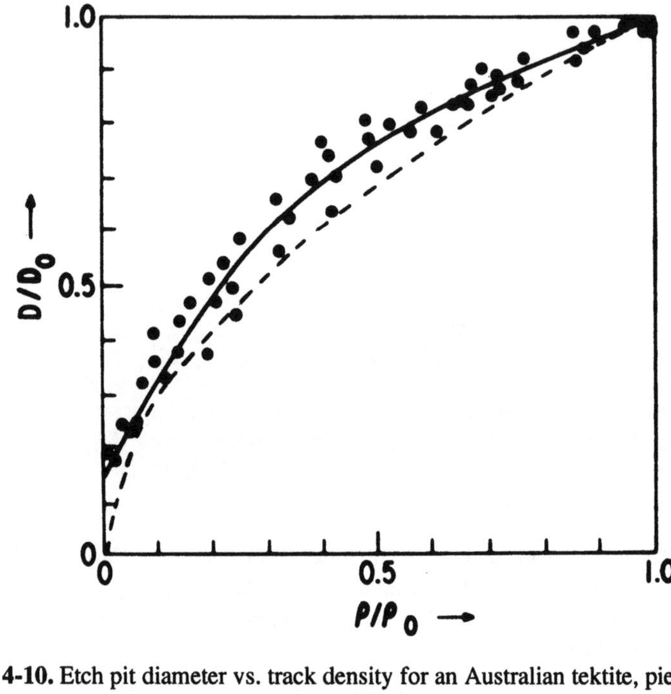

Figure 4-10. Etch pit diameter vs. track density for an Australian tektite, pieces of which were heated in the laboratory (Storzer and Wagner, 1969). The mean track diameter in each annealed sample is plotted against the track density for that sample; the data are relative to the average diameter and track density for an unannealed sample. The solid curve is smoothed through the points. The dashed curve was calculated using parameters measured on a synthetic tektite glass furnished by D. R. Chapman (FH72).

The annealing relation in Figure 4-10 is remarkable in that temperature and time are not explicitly involved. The figure implies that time and temperature are interchangeable in their effects on track size and therefore on track density, whose connection with size is purely geometrical (FH72). Westgate and co-workers (Westgate, 1988, 1989; Sandhu and Westgate, 1995) have developed this technique further.

Dating Thermal Events

Virtually all fission track dating gives the times of thermal events. Usually the event is the original cooling down of a mineral that formed from a hot liquid or solid. Below some critical temperature range all tracks are locked in, and the age is that of the original cooling through that temperature range.

Where subsequent events erase tracks, those events leave their marks, which are decipherable if the events are simple enough. Table 4-2 only shows that effects exist but does not explain them quantitatively.

One example of a simply interpretable thermal event is the case of the ages of Cathedral Peak granite shown in Figure 4-11 (Calk and Naeser, 1973). Here an 80-M.y.-old rock was intruded by molten lava 10 M.y. ago. Samples far from the contact show the higher age; those close to it give the lower, heat-affected age of the intrusion. And the transition region at a 3 – 10-m distance shows, first, where partial track retention occurred, and second, that the mineral sphene retains tracks to higher temperatures than does apatite.

Indeed, one way to measure the ancient temperature at a given time is to measure the track age of a suite of minerals in a rock (where the set has a range of track-annealing temperatures). If the most retentive have the highest age, and the progressively less retentive ones lower and lower ages, points on a cooling curve can be inferred. Price, Walker, and I used this sort of reasoning and more exotic fission tracks (from ^{244}Pu, which was present 4,600 M.y. ago) to infer that the meteorite Toluca cooled at a rate of ~1°C/M.y. through the temperature range 900K to 500K at the beginning of the solar system (Fleischer *et al.*, 1968b).

The Power of Track Length Distributions

There is far more information stored in heat-affected minerals than I have described so far, information of great help for geology in general and oil-exploration in particular. As mentioned earlier, tracks shorten as a result of heating. If a mineral was formed at time T_0 ago and heated so as to shorten tracks at T_1 ago, it will now contain a fraction $(T_0-T_1)/T_0$ of shortened tracks, a value which is derivable by measuring the resultant two peaks in the length distribution of tracks – the second peak being the full-length tracks that formed since T_1. Since the time and temperature of the heating affects how much the tracks are shortened, both the age when heating occurred and the amount of heating are recorded.

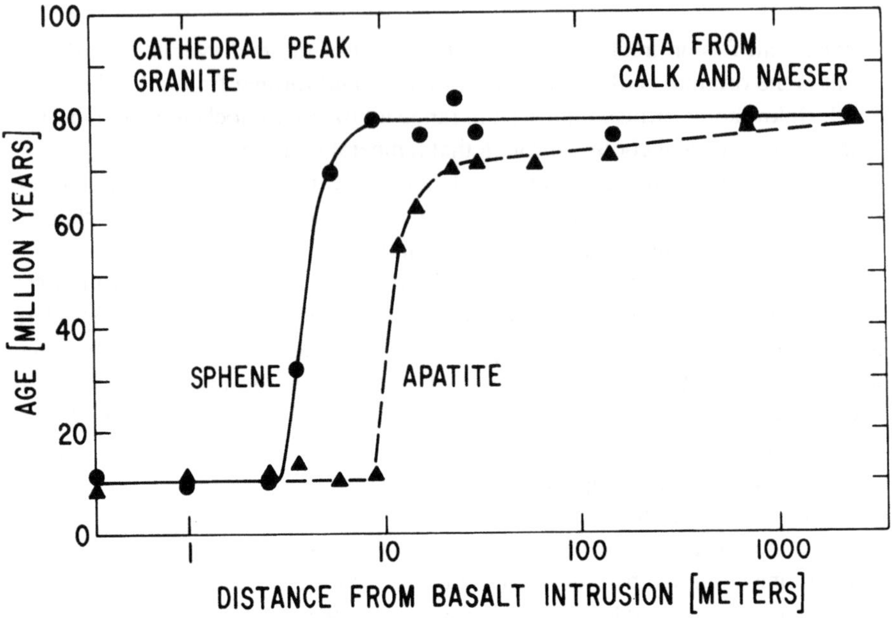

Figure 4-11. Effect of an igneous intrusion on fission-track ages. Ages are lowered where heating from the intrusion of a basalt plug into granite at Cathedral Peak, Yosemite National Park, caused track fading in apatite and sphene. The more retentive crystals (sphene) allowed fading only up to 10 meters from the contact as opposed to >100 meters for the less retentive mineral apatite (data of Calk and Naeser, 1973).

Similarly, if instead of the discrete two-step cooling and later brief reheating, the history includes a long time at a temperature where track shortening via partial annealing occurs, the consequence is not two characteristic track lengths but a continuous distribution of lengths. From measurements of the effects in the laboratory of time and temperature on tracks, the quantitative influence of longer times at lower temperatures can be inferred. Given this information, the effects of any arbitrary thermal history can be modeled and those results compared to actual observed length distributions. Such histories may include a progressively uplifted and eroded (and therefore slowly cooled) sequence or one of progressive burial (and therefore slow heating). Possibilities abound.

Simple histories can give simply recognized length distributions. Complex histories can not be understood uniquely using a single mineral sample, but samples from multiple depths or multiple minerals allow more details to be deciphered – more parameters being required to fit progressively more intricate situations. Such applications are plentiful. Two examples of the sorts of inferences that have been made are first that certain sediments in the Catskill region of New York State were cooled rapidly from above 110°C 130 M.y. ago, and indicate uplift and erosion of 3 to 4 km (Miller and Duddy, 1989) and second that minerals from Himalayian source rocks that are now found in the Bengal fan in India cooled "rapidly" (at more than 15°C/M.y.) as a result of "denudation rates of hundreds of meters per million years" (Corrigan and Crowley, 1992).

Track Length Measurement

To the beginner it is not obvious how track length distributions are best measured. Our original way was to polish a surface, etch it, and measure the lengths of the etched tracks that are intersected at random. If the original unetched tracks were all of etchable length R, all lengths from zero to R will be seen equally. Disentangling a mixture of R values is possible but difficult, and resolution is low. A better technique is track-in-track (TINT) etching (Lal, 1969). If a track that crosses the etched surface passes close to one that is entirely within the solid, etching can proceed from the first track along the buried one toward both ends, and thus its full etchable length is revealed (Figure 4-12). Most conveniently a sample is irradiated at perpendicular incidence by fission fragments, usually from a ^{252}Cf source, to allow many subsurface tracks to be intersected during etching. These will include ones that are nearly parallel to the original polished surface. They are measured in a microscope, with little obstruction from the dot-like appearance of the artificial, head-on tracks.

COMMERCIALIZATION

Because of the special ability of fission-track dating to describe thermal history, interest has intensified progressively. For example, the number of specialists who have contributed to the every-four-years meetings devoted to track dating rose from 50 in 1980 to 104 in 1984, 182 in 1988, and 286 in 1992.

And businesses have begun to serve the geoscience and geoengineering communities. The first of these was Geotrack International – begun at the University of Melbourne and established as an independent business in 1987. The Kyoto Fission-Track Company, Eurotrack at University College in London, and Donalick Analytical in Texas are three other such firms. It is now possible to submit a rock, have individual minerals separated, dated using track-length distributions, and interpretations provided. The scientific importance to geologists is obvious; less so is the role of track dating in exploration for oil.

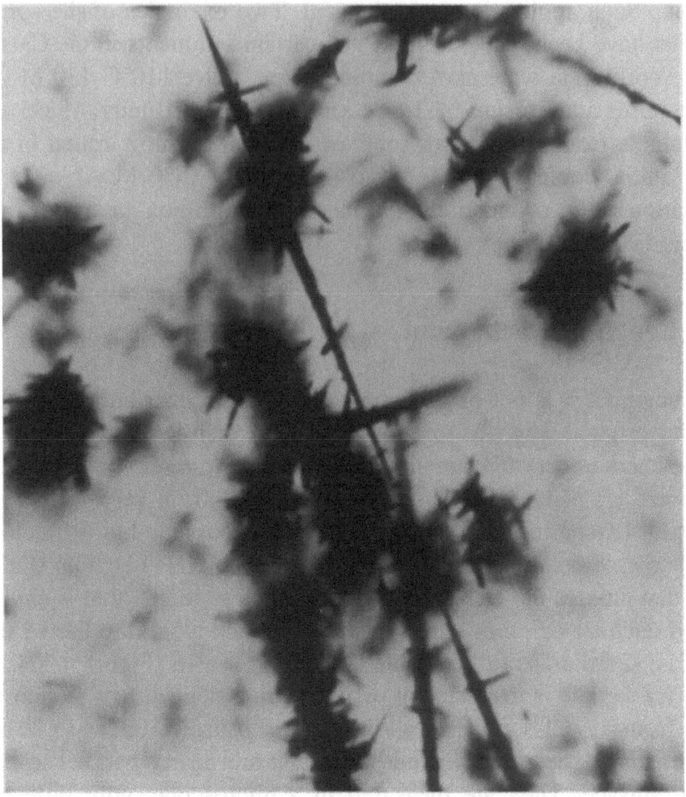

Figure 4-12. Etched track-in-tracks (TINTs) in a meteorite. Etching along the long track allows intersecting tracks to be etched in both directions to display their full etchable lengths (courtesy of D. Lal).

OIL EXPLORATION AND THERMAL HISTORY

Petroleum and natural gas are formed in the earth under certain specialized thermal conditions, and they are destroyed under others.

Formation of oil from organic molecules in the earth (*oil maturation*) occurs in the temperature range 60°C to 125°C and natural gas (primarily methane) up to 225°C (Hood *et al.*, 1975). These temperatures are ones to which the mineral apatite is specially suited, since tracks are heat-affected but not erased by long-term exposure to temperatures up to ~120°C (Naeser and Faul, 1969). Recent work by Carlson (1990), Crowley *et al.* (1991), and Laslett *et al.* (1987) give more complete results.

If mature oil migrates into a reservoir, its continued existence can be threatened by further heating, which may dissociate the oil and hence destroy a petroleum field. If preliminary sampling shows that such heating had occurred, vast wasted effort and cost in further exploratory drilling would be prevented.

A second geological factor in finding accessible oil is the need for a trap or reservoir in which maturing oil can concentrate. If dating can show that a potential reservoir formed only after oil maturation, it is a poor candidate for drilling. And the expense of drilling deep holes, which in places penetrate to depths of 7,500 m or more, is so immense (up to $100 million) that costs of $1,000 (±50%) for dating and thermal analysis of individual samples are thought of as relatively trivial investments. Fisher (1964) has collected data showing that drilling costs in the United States increase exponentially with depth, such that a single 7,500 m hole today typically might cost $20 million.

Such savings are the major economic value of thermal analysis by fission-track dating. Put differently, thermal dating can increase the probability of success in selecting a site for extensive drilling so that the average hole that is drilled is more productive.

One example comes from Atlantic Richfield's work in Jameson Land, eastern Greenland. Both fieldwork and apatite-track dating were done, with the track work ultimately concluding that Jameson Land basin was "overmature due to Mesozoic burial and Tertiary magmatism" (Bergman *et al.*, 1990). Exploration there has ended.

FURTHER READINGS ON TRACK DATING

Chapters on track dating:

FH72 *Fission Track Dating: Techniques and Problems*, R L. Fleischer and H. R. Hart, Jr., pages 135–170 in *Calibration of Hominoid Evolution*, W. W. Bishop and J. A. Miller, eds., Scottish Academic Press, Edinburgh, 1972.

FPW75 Chapter 4 in *Nuclear Tracks in Solids*, R. L. Fleischer, P. B. Price, and R. M. Walker, Univ. of California Press, Berkeley, 1975, pages 159 to 231.

A book on track dating:

WH92 *Fission-Track Dating*, G. A. Wagner and P. van den Haute, Enke, Stuttgart, 1992, 285 pages.

REFERENCES

H. Baumhauer (1894), *Die Resultante der Aetznmethode*, Leipzig (see WH92, p. xii).

S. C. Bergman, F. H. Henk, and K. Hansen (1990), "East Greenland Apatite Fission Track and Magmatic Rock Record of North Atlantic Spreading and Hotspot Events," Seventh Int. Conf. on Geochronology, Cosmochronology, and Isotope Geology, Canberra, Sept.

R. H. Brill, R. L. Fleischer, P. B. Price, and R. M. Walker (1964), "The Fission Track Dating of Man-Made Glasses: Part I, Preliminary Results," *J. Glass Studies* **6**, 151–155.

L. C. Calk and C. W. Naeser (1973), "The Thermal Effect of a Basalt Intrusion on Fission Tracks in Quartz Monzonite," *J. Geol.* **81**, 189–198.

W. D. Carlson (1990), "Mechanisms and Kinetics of Apatite Fission-Track Annealing," *Am. Mineral.* **75**, 1120–1139.

D. Corrigan and K. D. Crowley (1992), "Unroofing of the Himalayas: A View from Apatite Fission-Track Analysis of Benegal Fan Sediment," *Geophys. Res. Lett.* **19**, 2345–2348.

K. D. Crowley, M. Cameron, and B. J. McPherson (1990), "Annealing of Etchable Fission-Track Damage in F-, OH-, Cl-, and Sr-Apatite--1. Systematics and Preliminary Interpretations" [abs.], *Nuclear Tracks and Radiation Measurements* **17**, 409–410.

K. D. Crowley, M. Cameron, and R. L. Schaefer (1991), "Experimental Studies of Annealing of Etched Fission Tracks in Fluorapatite," *Geochim. Cosmochim. Acta* **55**, 1449–1465.

F. M. Fisher (1964), *Supply and Costs in the U. S. Petroleum Industry*, Johns Hopkins Press, Baltimore.

R. L. Fleischer (1993), "Remembrances of Tracks Past," *On Track* **3**, 6–7.

R. L. Fleischer, P. B. Price, E. M. Symes, and D. S. Miller (1964), "Fission Track Ages and Track-Annealing Behavior of Some Micas," *Science* **143**, 349–351.

R. L. Fleischer, P. B. Price, R. M. Walker, and L. S. B. Leakey (1965a), "Fission Track Dating of Bed I, Olduvai Gorge," Science **148**, 72–74.

R. L. Fleischer, P. B. Price, R. M. Walker, and L. S. B. Leakey (1965b), "Fission Track Dating of a Mesolithic Knife," *Nature* **205**, 1138.

R. L. Fleischer, J. R. M. Viertl, P. B. Price, and F. Aumento (1968a), "Mid-Atlantic Ridge: Age and Spreading Rates," *Science* **161**, 1339–1342.

R. L. Fleischer, P. B. Price, and R. M. Walker (1968b), "Identification of Pu^{244} Fission Tracks and the Cooling of the Parent Body of the Toluca Meteorite," *Geochim. Cosmochim. Acta* **32**, 21–31.

R. L. Fleischer, J. R. M. Viertl, P. B. Price, and F. Aumento (1971), "A Chronological Test of Ocean Bottom Spreading in the North Atlantic," *Radiat. Effects* **11**, 193–194.

R. L. Fleischer and P. B. Price (1964a), "Glass Dating by Fission Fragment Tracks," *J. Geophys. Res.* **69**, 331–339.

R. L. Fleischer and P. B. Price (1964b), "Fission Track Evidence for the Simultaneous Origin of Tektites and Other Natural Glasses," *Geochim. Cosmochim. Acta* **28**, 755–760.

G. N. Flerov and K. A. Petrjak (1940), "Spontaneous Fission of Uranium," *Phys. Rev.* **58**, 89. (more details in Doklady [Comptes Rendus] **28**, 500 [1940] in Russian.)

P. F. Green, I. R. Duddy, A. J. W. Gleadow, and J. F. Lovering (1989), "Apatite Fission-Track Analysis as a Paleotemperature Indicator for Hydrocarbon Exploration," in N. D. Naeser and T. H. McCulloh, eds., *Thermal History of Sedimentary Basins—Methods and Case Histories.* New York, Springer-Verlag, pp. 181–195.

A. Hood, C. C. M. Gutjahr, and R. L. Heacock (1975), "Organic Metamorphism and the Generation of Petroleum," *Bull. Am. Assoc. Petr. Geol.* **59**, 986–996.

D. Lal (1969), "Recent Advances in the Study of Fossil Tracks in Meteorites Due to Heavy Nuclei of the Cosmic Radiation," *Space Sci. Rev.* **9**, 623–650.

G. M. Laslett, P. F. Green, I. R. Duddy, and A. J. W. Gleadow (1987), "Thermal Annealing of Fission Tracks in Apatite—2. A Quatitative Analysis," *Chem. Geol.* **65**, 1–13.

L. S. B. Leakey, J. F. Evernden, and G. H. Curtis (1961), "Age of Bed I, Olduvai Gorge, Tanganyika," *Nature* **191**, 478–479.

M. Maurette (1976), "Fossil Nuclear Reactors," Annu. Rev. Nuc. Sci. **26**, 319–350.

M. Maurette, P. Pellas, and R. M. Walker (1964), "Etude des Traces de Fission Fossiles dans le Mica," *Bull. Soc. Franc. Min. Crist.* **87**, 6–17.

D. S. Miller and I. R. Duddy (1989), "Early Cretaceous Uplift and Erosion of the Northern Appalachian New York, Based on Apatite Fission Track Analysis," *Earth Planet. Sci. Lett.* **93**, 35–49.

C. W. Naeser and H. Faul (1969), "Fission-Track Annealing in Apatite and Sphene," *J. Geophys. Res.* **74**, 705–710.

J. A. O'Keefe (1963), *Tektites*, Univ. of Chicago Press, Chicago.

P. B. Price and R. M. Walker (1962), "Observation of Fossil Particle Tracks in Natural Micas," *Nature* **196**, 732–734.

P. B. Price and R. M. Walker (1963), "Fossil Tracks of Charged Particles in Mica and the Age of Minerals," *J. Geophys. Res.* **68**, 4847–4862.

A. S. Sandhu and J. A. Westgate (1995), "The Correlation Between Reduction in Fission-Track Diameter and Areal Track Density in Volcanic Glass Shards and Its Application in Dating Tephra Beds," *Earth Planet. Sci. Lett.* **131**, 289–299.

D. Storzer and M. Sélo (1976), Uranium Contents and Fission Track Ages of Some Basalts from the FAMOUS Area, *Bull. Soc. Geol. Fr.* **18**, 807–810.

D. Storzer and G. A. Wagner (1969), "Correction of Thermally Lowered Fission Track Ages of Tektites," *Earth Planet. Sci. Lett.* **5**, 463–468.

J. A. Westgate (1988), "Isothermal Plateau Fission-Track Age of the Late Pleistocene Old Crow Tephra, Alaska," *Geophys. Res. Lett.* **15**, 376–379.

J. A. Westgate (1989), "Isothermal Plateau Fission-Track Ages of Hydrated Glass Shards from Silicic Tephra Beds," *Earth Planet. Sci. Lett.* **95**, 226–234.

Chapter 5
COSMIC RAYS

Doubt thou the stars are fire
W. Shakespeare, in *Hamlet*

The space above the atmosphere of the earth – between the planets and the stars – is thought by many people only to be empty space. Yet it contains and is traversed by an immense diversity of radiation, both electromagnetic and nuclear. Occasional dust particles are also present. The radiation carries information about distant bodies in the cosmos, and its complexity delivers elaborate clues to the nature of astrophysical processes that have emitted vast quantities of energy. Examples of these sources are the surface of our sun, supernova explosions, pulsars, and red giant stars.

The atomic nuclei that are accelerated from such sources are of particular interest to people who use track etching, because solid track detectors often are uniquely efficient instruments to record and individually distinguish these cosmic ray particles. And where natural track detectors have recorded ancient fluxes of particles, these fossil tracks in bodies such as meteorites and lunar samples help us learn about their history as well as the environments that they have so long endured.

FOSSIL TRACKS OF COSMIC RAYS

The beginnings of using etched tracks to observe cosmic rays rested on Walker's partially naive hope to detect cosmic ray particles that entered mica crystals on the surface of the moon. When finally lunar samples were returned, such tracks were in fact abundant in the Apollo 11 samples, and they were observed by many groups of track workers (Proceedings, 1970) – but not in mica, which we later learned was unknown in extraterrestrial material.

The cosmic ray tracks in Figure 5-1, a small crystal of lunar augite ($CaMgSi_2O_6$), allow us to view the past irradiation. How do I know what caused these tracks? As will be amplified later, their length and orientation distributions are remarkably diagnostic. Fission tracks are oriented at random and the range of the particles (~10 μm) is strictly limited by the energy release in fission (~180

MeV). Cosmic ray tracks can be both shorter and far longer, and usually they have a preferred orientation along the direction to the nearest surface of the meteorite or lunar sample. The very few tracks in Figure 5-1 give a strong hint of preferred track orientation, even though the number of tracks is far too few to establish the anisotropy quantitatively.

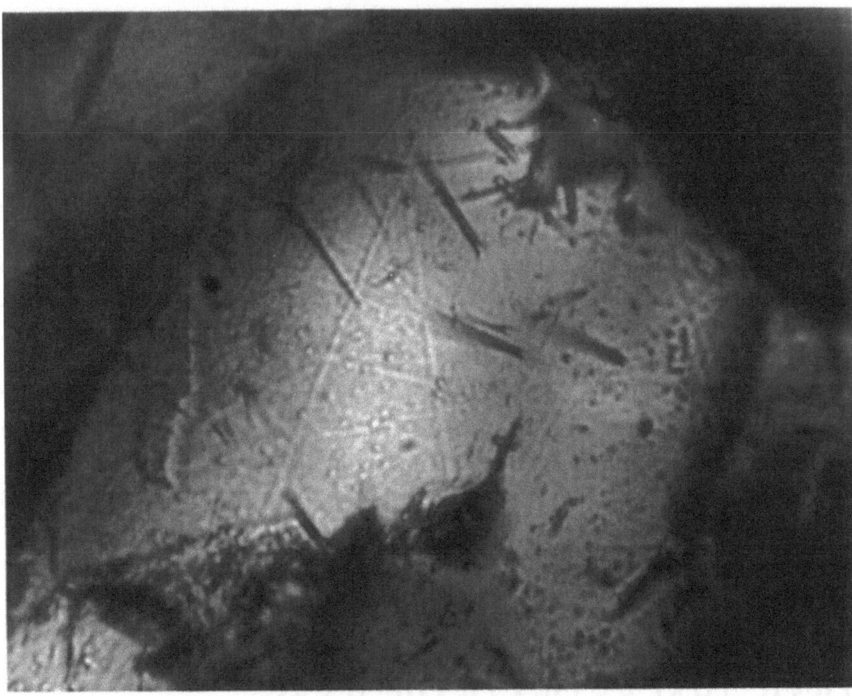

Figure 5-1. Cosmic ray tracks etched in a small pyroxene grain from partway into lunar rock 10017. The preferential orientation of these tracks, the longer ones, is evident. The many dot-like shallow, short tracks are from spallation recoil events – proton-induced nuclear fragmentations of atomic nuclei (Fleischer *et al.*, 1970a).

Heavy Cosmic Rays

The cosmic rays in space mostly consist of atomic nuclei, the most abundant ones being the lightest, protons; and the second most abundant, He nuclei, the second-lightest element. Heavier nuclei, although rarer, are now known through most of the periodic chart of elements, but as of 1964 the heaviest known were

Fe and its near neighbors in atomic number (Freier *et al.*, 1948). Nuclear track etching has played a major role in the expansion of our knowledge about the composition of the heaviest cosmic rays, and the relative abundances are evidence as to how and where these nuclei came to be accelerated into interstellar space.

The first reported etched tracks associated with cosmic rays were seen in the meteorite Pavlodar by Maurette *et al.*, (1964) and attributed to cosmic-ray-induced nuclear reactions with atomic nuclei in the meteorite. By the next year (Fleischer *et al.* 1965), we had seen large numbers of tracks in 14 minerals from 33 meteorites. Later we clarified the types of tracks that can occur, suggested simple criteria for identification, and showed that the most commonly observed type of track in meteorites is created directly by heavy cosmic ray particles – nuclei having masses close to those of the iron isotopes (Fleischer *et al.*, 1967a,b).

Meteorites are well-known objects from space that fall on the earth from time to time, often with a spectacular blaze of light when they drop at night. They are strongly heated and ablated as they are slowed in the upper atmosphere, with the result that in some cases only a small fraction of the original mass reaches the earth. Each is named after a place near where it fell, the Schenectady meteorite shown in Figure 5-2 being an example. The final mass here, 283 gm, is about 1/300 of its pre-atmospheric value (Fleischer *et al.*, 1970b). This object spent a measured 31 M.y. in space and hence it accumulated a long record of the incidence of cosmic rays. Prior to 31 M.y. ago Schenectady presumably was part of a much larger body that shielded it.

This meteorite did in fact arrive at night and was observed widely across New York State and north into Canada. The only useful information from such observations is the possibility of inferring the azimuthal direction from which it arrived. If a viewer sees the light transit from right to left, the 180° sector on the left is eliminated. Assembling such data from many different vantage points around its landing point should narrow its possible direction. In this case the 11 observations that were collected were more than enough; they eliminated *all* directions of arrival, all but one corridor rejected by multiple observers. A consistent result, that the object came from the east, could be inferred by eliminating one particular observer. He happened to be the one trained scientist in the group; but he was known to be flat on his back at the time, enjoying a beer party.

As an illustration as to how identification is possible for different types of tracks, consider tracks from spontaneous fission and cosmic rays. Spontaneous fission is random in direction, is expected to occur in those minerals where uranium is present, and will give tracks of nearly constant length. Cosmic rays come from outside, will therefore be attenuated with depth, will be nearly equally abundant in adjacent minerals of differing crystal type, and (it turns out) will have an informative wide spectrum of lengths. Figures 5-3 and 5-4 show

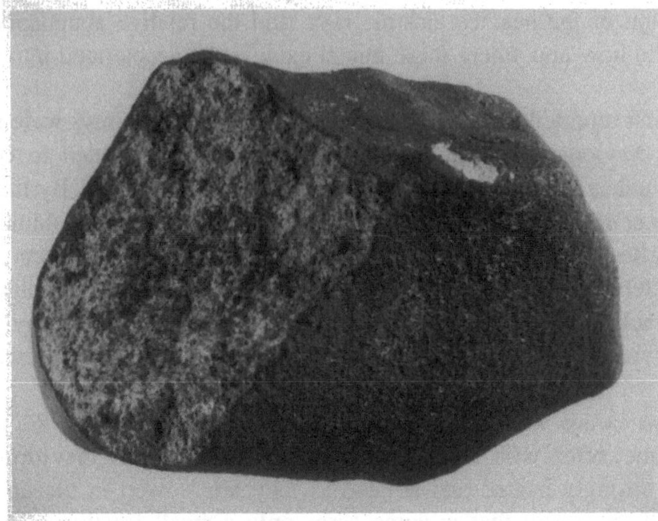

Figure 5-2. The Schenectady meteorite. The fresh (white) surface was produced on impact; the darker fracture surface at left was produced during descent through the Earth's atmosphere and is dotted with dark globs of asphalt roofing material, from hitting the roof of a house. The owner, who was watching, television at the time, said that its impact "sounded like a firecracker going off" in his attic. He then returned to watching TV. The largest dimension, as shown, is 6.9 cm (Fleischer *et al.*, 1970b).

examples. Estherville was sectioned and tracks counted in minerals from various positions. Figure 5-3 shows variations by a factor of 60 with position, the lower left surface being inferred to be closest to the pre-atmospheric surface. Figure 5-4 shows track length distributions on polished surfaces for Estherville and another meteorite, Odessa, with very different tracks. As described in Chapter 4, fission tracks of a single original length are cut randomly by a surface into equal numbers in each length interval up to the original value – as seen in the Odessa data. The cosmic ray tracks in Estherville are totally different and include a small fraction of much longer tracks.

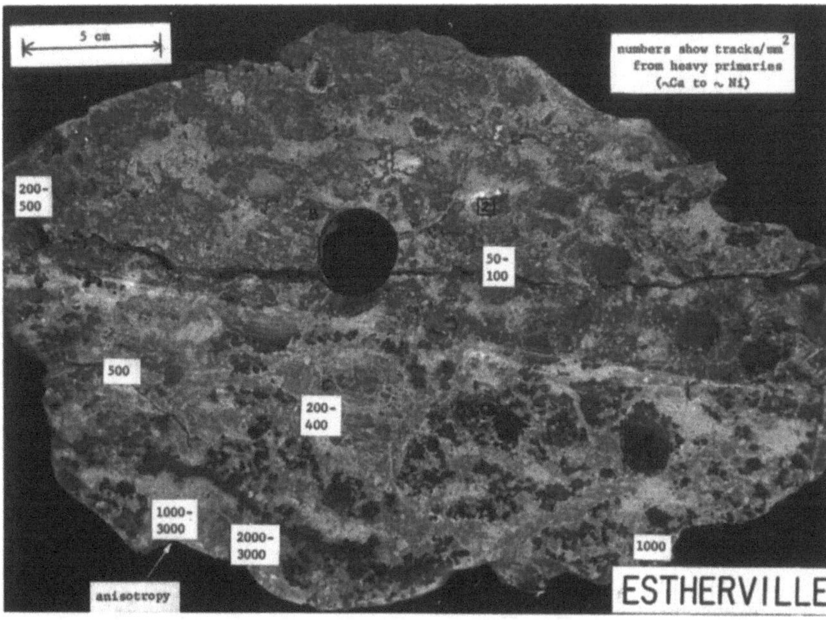

Figure 5-3. Photograph showing locations and track densities in hypersthene crystals taken from a slab cut from the Estherville meteorite. The arrow indicates the preferred direction of the heavy primary cosmic rays (Fleischer *et al.*, 1967b).

Elemental Abundances and Track Lengths

The track lengths in minerals are measures of the elemental distribution, the long tracks in Figure 5-4 for Estherville being from heavier cosmic rays than were previously known; they are made by trans-iron nuclei. To see how this connection comes about, we use Figure 5-5, an expanded version of Figure 1-6 (right), which was used to explain the track formation threshold. The experimental points are for Lexan polycarbonate and show a discrete threshold at ~7 on the ionization scale. Meteoritic minerals are indicated by a ten-times-higher threshold. Argon just fails to reach this threshold, and therefore a cosmic ray

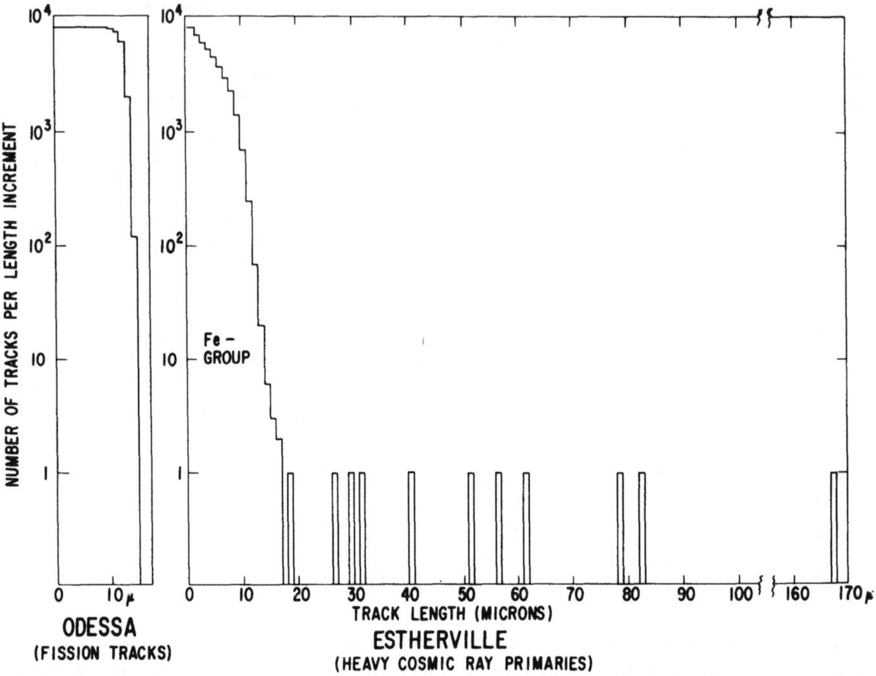

Figure 5-4. Track length distribution observed in Estherville compared to the typical fission track-length distribution observed in the iron meteorite Odessa. The existence of tracks of up to 160 μ in length implies the presence of cosmic-ray nuclei of charge considerably heavier than that of iron (Fleischer *et al.*, 1967b).

argon nucleus does not form a track, but Fe will do so below about 1.5 MeV/amu (atomic mass unit) and Kr below about 5 MeV/amu. Range (distance till the particle comes to rest) increases with energy per mass, and the 5-MeV/amu Kr has about 3 times the range of a 1.5-MeV/amu Fe. Hence, as fast-moving Kr and Fe slow down inside a meteorite, once they cross threshold they begin to form tracks; and the Kr track is 3 times as long as that from Fe. Pb at its threshold of about 50 MeV/amu will have an etchable range that is more than 10 times longer than that of the Kr.

Subsequent work in several laboratories has revealed tracks that extend up to 1 mm both in meteorites and in minerals from the lunar surface, strong indication that nuclei that are far heavier than iron were present in the historic cosmic radiation. Walker's discussion in Chapter 6 of FPW75 covers that work and some of the subtleties that need to be included in untangling the fossil records of the mass distribution of heavy cosmic rays. What is established is that most of the elements in the periodic chart are present in cosmic rays at least through Pb and Bi.

We now turn to modern-day cosmic rays.

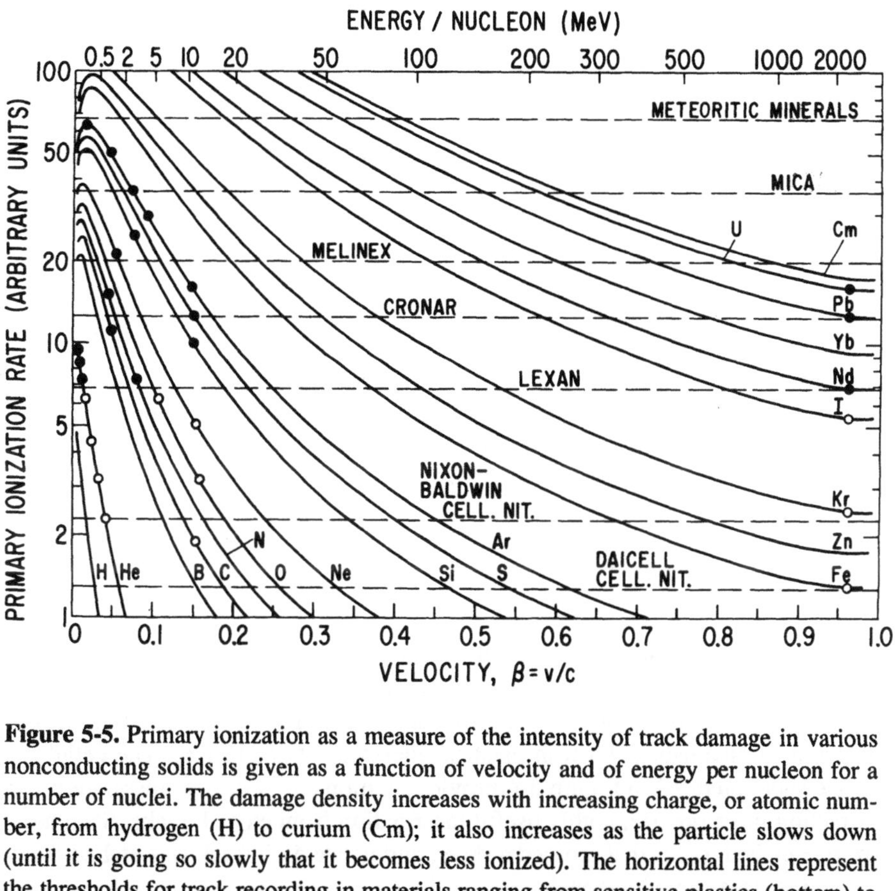

Figure 5-5. Primary ionization as a measure of the intensity of track damage in various nonconducting solids is given as a function of velocity and of energy per nucleon for a number of nuclei. The damage density increases with increasing charge, or atomic number, from hydrogen (H) to curium (Cm); it also increases as the particle slows down (until it is going so slowly that it becomes less ionized). The horizontal lines represent the thresholds for track recording in materials ranging from sensitive plastics (bottom) to typical constituents of meteorities (top). The experimental points for accelerator ions in Lexan polycarbonate are given as open circles for zero registration and as filled circles for 100% registration.

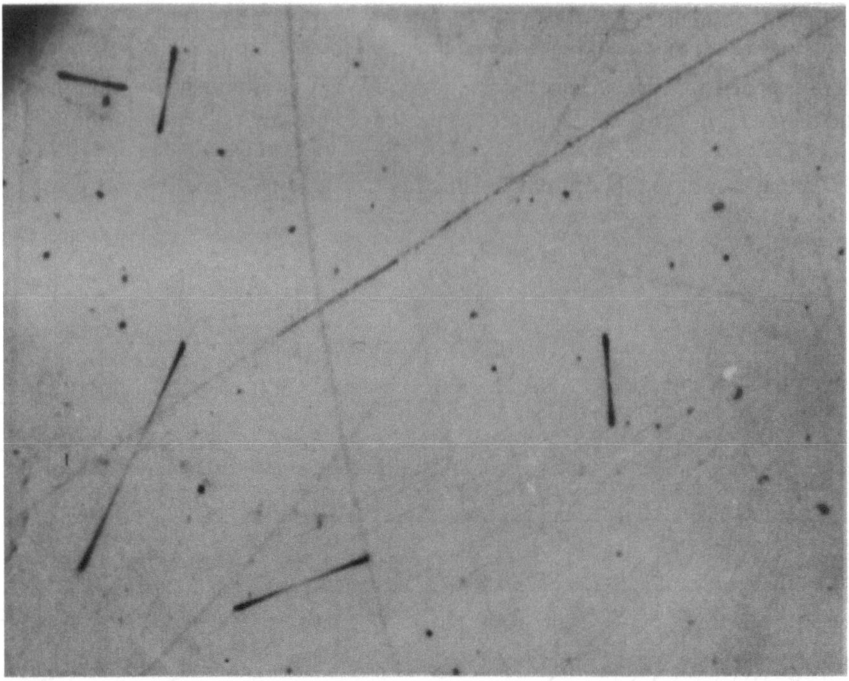

Figure 5-6. Cosmic ray tracks etched from both sides through a 250-μm-thick disk of cellulose nitrate exposed for 7 days in a polar-orbiting satellite. Different tracks have different track etching rates as shown by the spread of tapers. The longest track is 600 μm in projected length (Fleischer *et al.*, 1967c).

PARTICLE IDENTIFICATION

Over the same period that the fossil tracks were being studied, we were also using high-altitude balloons and space flights to observe present-day cosmic rays. The data helped lead us to the ionization criterion for registration thresholds and to establish its level in plastics (Fleischer *et al.*, 1967c). The cellulose nitrate sheet pictured in Figure 5-6 provided a moment of critical inspiration for extracting more precise, highly useful information from etched tracks. Seen here are five etched tracks that crossed a sheet, and hence each track was etched from the two surfaces simultaneously.

One morning I found Price puzzling over this photo, believing correctly that there was a message to be deciphered. Through the earlier chapters I usually have tacitly implied that the velocity of etching along a track does not vary. What was surprising in Figure 5-6 is that different tracks have different shapes because of different rates of taper along their lengths. The three that are nearly the same length have very different widths midway along their lengths. Remembering Figure 1-3, the taper is given by the ratio of the general rate of etching (which is constant) to the track rate, and therefore *the different tracks have different preferential rates of attack,* v_T.

Price shortly recognized that individual particles might be identified if these effects reproduce reliably. Consider the Lexan data in Figure 5-5. What the tracks in Figure 5-6 imply is that the solid dots each correspond to a different etching rate. Further up in the graph, where ionization is higher, the damage is more intense, and the chemical attack rate is correspondingly increased.Once calibrated, each etching rate corresponds to an ionization level in Figure 5-5, so that if we can also determine the energy (or velocity) for an unknown particle, we can plot a position on the figure, and that location identifies what nucleus we have. Because a given energy particle has a predictable range, the energy is determined by finding where a particle comes to rest.

Thus began a new era in the science of etched tracks, the ability to extract quantitative information about individual particles. We shortly evolved two procedures, the first of which (Price *et al.*, 1967), the multiple-sheet method, is illustrated in Figure 5-7. A particle that crosses three detector sheets comes to rest in sheet 4, and all 4 are then etched. Since ionization increases downward, the etched cone-shaped tracks lengthen as sketched, and the final etched shape is test-tube-like, blunted where preferential etching ended at the site where the particle came to rest.

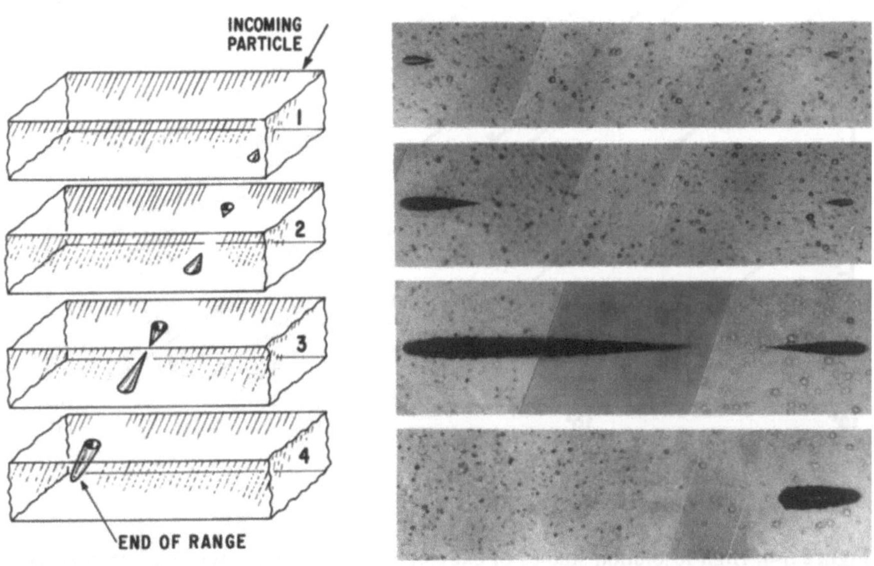

Figure 5-7. Etched portions of the track of a Ca cosmic ray slowing down in a stack of Lexan sheets. The rate of change of etched cone length with distance is a unique function of atomic number and mass. As is seen, the lengths of the etched cones increase from right to left and top to bottom through sheets 1, 2, and 3 as the velocity of the particle decreases in a diagnostic way until finally the particle stops in sheet 4 (Price and Fleischer, 1970).

The lengths of each of six cones give six values of the ionization, and the distance from each to the final rounded-out location gives the six ranges, providing in this case a sixfold redundancy that improves the quality of the measurement of this cosmic ray, a Ca ion. The power of this technique is shown by the fact that for particles accelerated in the laboratory, neighboring isotopes of boron can be clearly distinguished by their etched tracks.

In Figure 5-8, data for cosmic rays in a stack flown beneath a balloon illustrate the resolving power of the method and the redundancy that can be achieved by following nuclei through centimeters of multiple sheets (O'Sullivan *et al.*, 1971). Si and Fe are so abundant in the cosmic rays that they provide the internal calibration that is used to bootstrap identification all the way to U.

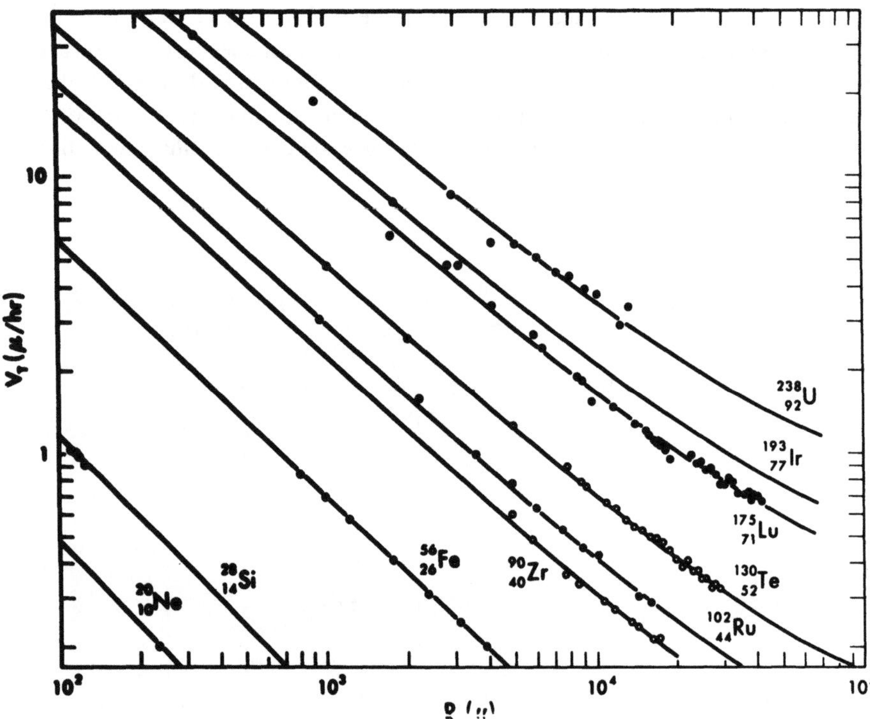

Figure 5-8. High-resolution studies of extremely heavy cosmic rays in a Lexan stack by the multiple-sheet method described by Figure 5-7. The Ne and Si data are averages of measurements of many tracks from accelerator bombardments. The Fe data represent the spread in measurements of about 50 cosmic rays that stopped in the stack. Data for other, less abundant elements are omitted for clarity. The data points for the six nuclei assigned charges of 40, 44, 52, 71, 77, and 92 give etch rate values at many positions along their trajectories in the stack (O'Sullivan *et al.*, 1971).

Alternatively the thick-detector technique shown in Figure 5-9 may be used (Fleischer *et al.*, 1970c). Once a slowing particle starts to produce a track, it may be etched continuously. Each point along the single etched track corresponds to a particular taper, θ, and hence a v_T value (for fixed, known v_G). The redundancy here is limited only by irregularities along the etched track that might cause θ not to vary smoothly.

But, as Figure 5-10 demonstrates, etching can proceed very smoothly indeed. These four cosmic rays encountered in a miscellaneous Lexan part in Apollo 14 are Ar, Fe, Ca, and Ti ions, clockwise from upper left (Fleischer *et al.*, 1973).

The same identification principles that are shown here for plastics apply equally to particle identification in many minerals (including biotite mica, pyroxenes, and feldspars) and a wide variety of inorganic glasses (PF71, FPW75).

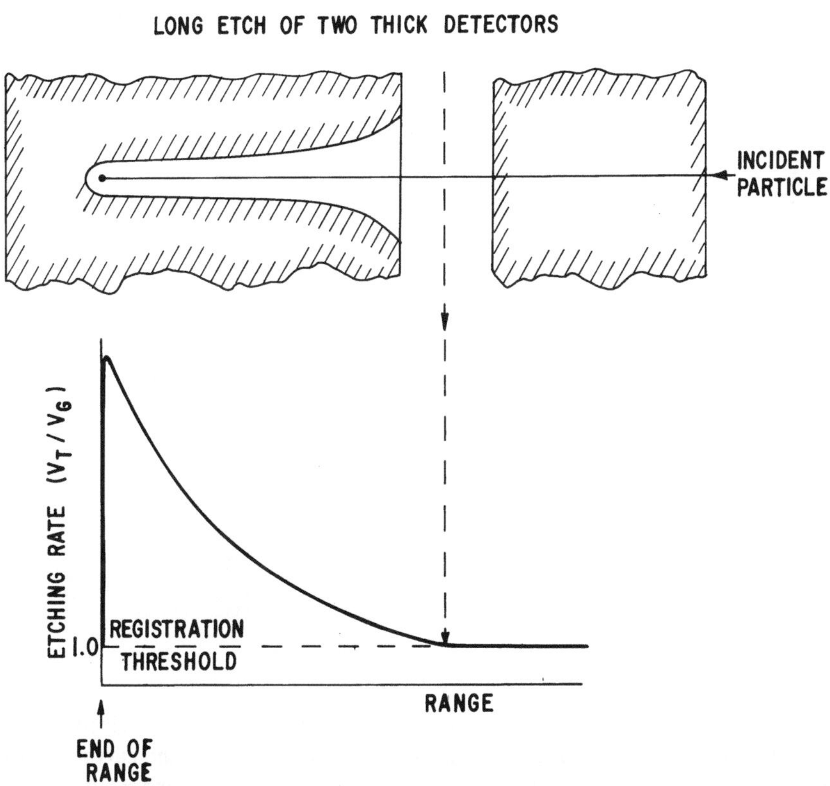

Figure 5-9. Thick-detector method of track identification. The taper as a function of position along the track is a measure of the ionization density, as expressed by preferential etching (v_T/v_G)

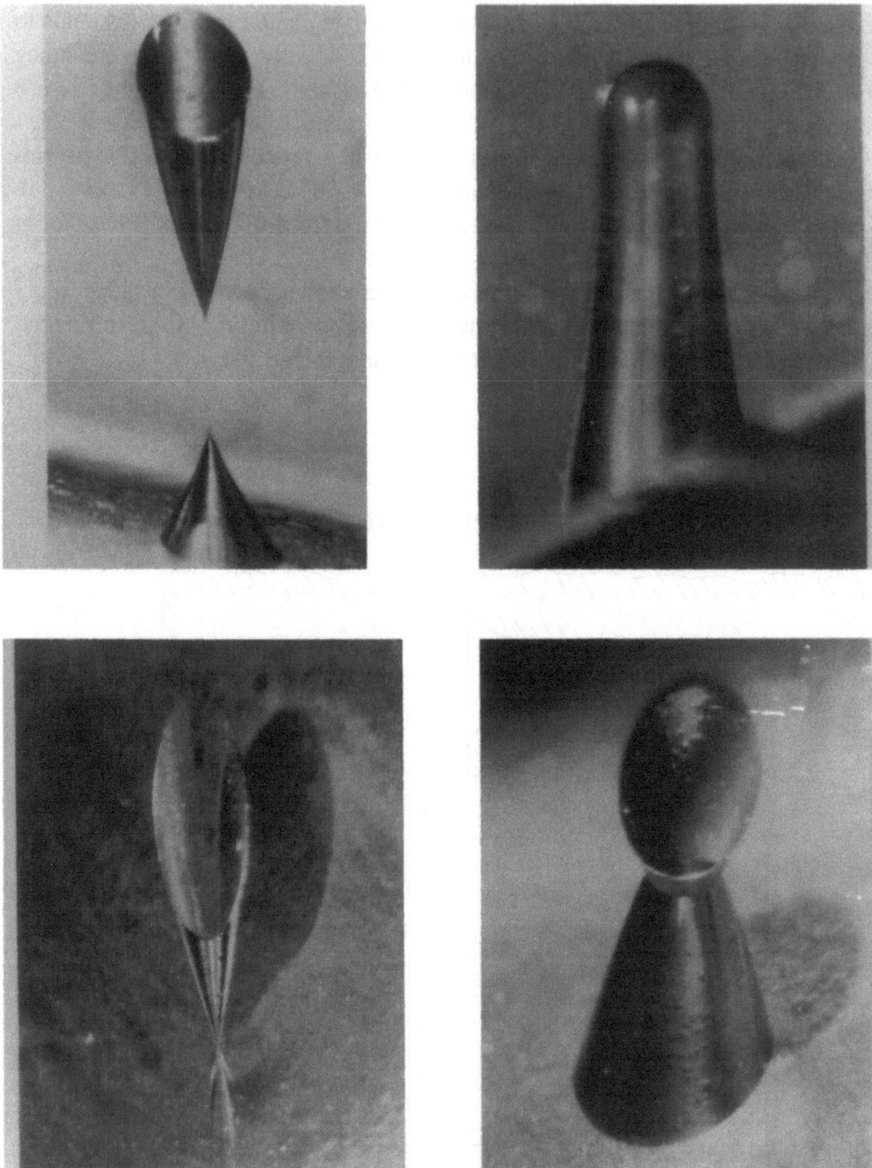

Figure 5-10. Etched tracks of cosmic rays that penetrated an Apollo 14 electrophoresis device that happened to be made of Lexan polycarbonate. The taper and curvature of the tracks of the particles allow them to be individually identified. From upper left clockwise they are argon, iron, calcium, and titanium ions (Fleischer *et al.*, 1973).

SUPERHEAVY COSMIC RAYS

Although there has been much use of track etching to measure abundances of cosmic rays, including elements of medium mass and most elements at moderate energy, the early excitement over new results came from the search to identify the heaviest elements that exist in cosmic rays. Those measurements are easiest for very fast particles.

Figure 5-5 tells us why high energy is preferred. At a velocity of 0.1 c (where c is the velocity of light) in Lexan, oxygen and all heavier elements will form tracks. Since lighter elements are generally more abundant, the occasional, rare very heavy nucleus will form a track in the midst of an intense background of the lighter nuclei.

At 0.9 c, however, only elements at least as heavy as the rare earths will form tracks, so that all recorded particles are interestingly heavy. There is a further simplification. The ionization rate changes little with energy in this region, with the result that throughout a thick stack of detectors, a given particle has etched cones that are very nearly identical; and thus their average length (or cone angle) identifies the element. And the particle need not be followed to rest.

Figure 5-11 is an example of a erbium or thulium nucleus encountered in a high-altitude balloon flight that was carried out close to the magnetic equator in order that the earth's magnetic field would turn away the unwanted, low-energy particles. The two detectors shown here, cellulose nitrate and Lexan polycarbonate, have very different thresholds. The Lexan sheets in the stack identify the particle; the cellulose nitrate provides a perforation, thus making location of tracks easy.

Industrial Help

It is a daunting task to scan 10 or 20 m^2 of detector at the magnification needed to see tracks like those in the Lexan pictured in Figure 5-11, but the holes through the cellulose nitrate simplified our task greatly. In what turned out to be a gift of technology to science, we learned that General Electric's Capacitor Department had been using an Ozalid copier as a quality-control device to scan dielectric material for undesired pinholes that might nucleate electrical breakdown in capacitors. Although Ozalid machines are now outdated and hard to find, they had a useful feature, the combination of ammonia and ammonia-sensitive paper to develop images. In our case, a detector sheet that might contain holes was run through the machine adjacent to a piece of the sensitive paper, ammonia blew through the holes, and it stained the paper locally – providing an image like a star map – pinpointing locations of holes. The corresponding etched tracks in the Lexan are then found by scanning tiny areas adjacent to the identified holes in the cellulose nitrate.

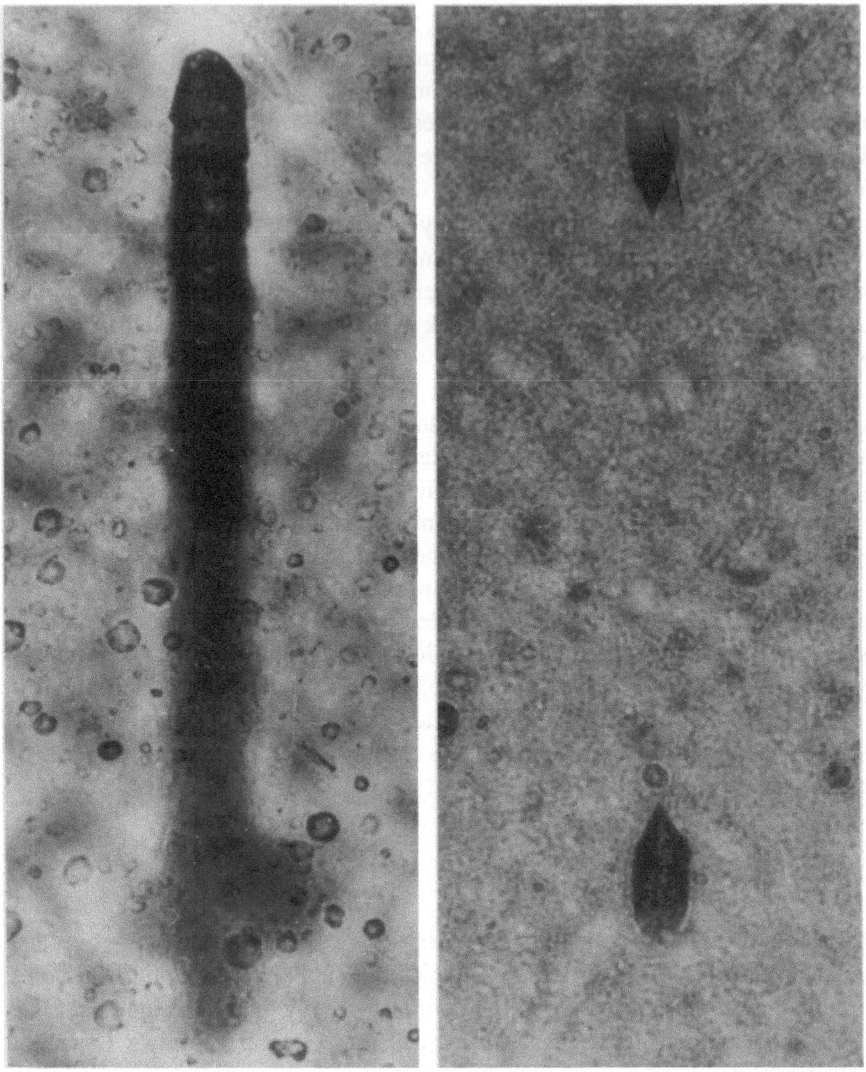

Figure 5-11. Tracks in two different plastic sheets left by a single particle whose ionization is equivalent to that expected of a charge 68 or 69 ion. Right: Track in Lexan polycarbonate. Left: Track in cellulose nitrate. The two sheets in which the tracks appear were part of a stack flown in a balloon at an altitude of 130,000 feet to observe heavy cosmic ray particles (Fleischer *et al.*, 1970d).

Observations

The finding of trans-iron cosmic ray tracks in meteorites ignited a race for observing heavy nuclei in the present-day cosmic rays. Professor Peter Fowler and co-workers at the University of Bristol used nuclear photographic film (*nuclear emulsions*) and brute-force scanning and reported that the two heaviest nuclei they saw were close to thorium, atomic number 90 (±4) (Fowler *et al.*, 1967). Our subsequent balloon flight (Blanford *et al.*, 1969) gave similar results with plastics; Fowler *et al.* (1970) found a nucleus with atomic number above 100; and finally Price and I combined techniques with Fowler, the enthusiastic, vigorous master of nuclear emulsions, using a detector array consisting of interleaved plastic detectors and emulsions (Price *et al.*, 1971a).

One of the results of this collaboration was the finding of one nucleus that the plastics implied was element 90 or 92 and Fowler's emulsion placed above 100. We were arguing the issue throughout a cosmic ray conference at Budapest in 1969, and the discussion extended into the midconference weekend at Lake Balaton, as we strolled past the cottage pictured in Figure 5-12. Its preferred, revised numbering appeared to convey the Hungarian vote to supplant 108 by 90. This choice turned out to be consistent with the interpretation that has prevailed.

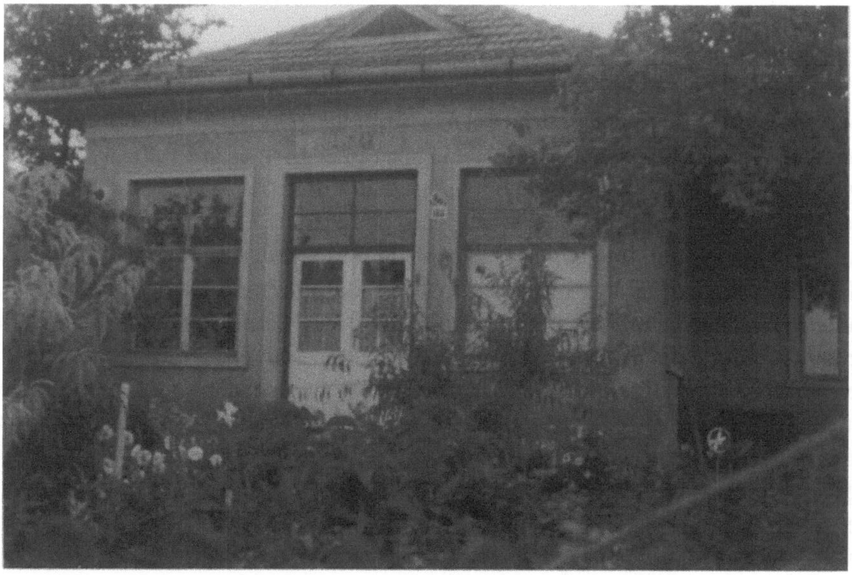

Figure 5-12. A cottage at Lake Balaton, Hungary where the number 90 has been chosen to replace 108. Is thorium more likely in the cosmic rays than element 108?

In more recent times Fowler turned from emulsions to plastic detectors, which are inexpensive and less environmentally sensitive to elevated temperature and low humidity. Incidentally, most of the balloon flights discussed were flown from the National Center for Atmospheric Research's Balloon Flight Station at Palestine, Texas. It is easy to say "we flew a load of detectors for 40 hours near the top of the atmosphere," but the actual job is interesting, laborious, and uncertain. Preparation required a couple of weeks constructing a lightweight platform to support the detectors, numbering, cutting to size, and stacking the sheets, and then enclosing them in thermal-control materials. Launches usually are delayed to await calm enough winds to avoid catastrophe. Then, if the balloon doesn't rip during take-off, comes a moment of quiet majesty as the vertical chain of the balloon, the parachute, suspension, and load silently lift into the dawn sky. The flight path and altitude were monitored; and when helium loss from the balloon was lowering it too far into the atmosphere, a remote signal was sent to separate the load, to descend by parachute. Our final flight with Fowler involved a vastly larger detector area than previously, enough for a good sample of the high atomic number cosmic rays. The primary electronic command failed to set loose the load, as did the back-up procedure, and the flight disappeared out over the Pacific Ocean.

Interestingly, a co-worker whom Fowler brought to Palestine was Moses (Roger T. Moses). There is no known biblical significance to this fact.

What Do Heavy Cosmic Rays Tell Us?

The cosmic rays are a sample of matter from distant locations and hence tell us about processes that are difficult to diagnose directly. The elements have long been thought to have been born in stars (Burbidge *et al.*, 1957), and those heavier than Fe come primarily from two types of sites that are rich in neutrons, which build heavier nuclides, one mass unit at a time. A class of stars known as red giants have neutrons in relatively low abundance. The neutrons are slowly added to nuclei, building heavier elements in the so called s-process (*s* for slow). The importance of adding neutrons slowly is that time for beta decay is allowed and no elements above 82 and 83, the lead group (Pb and Bi), are produced. In contrast, rare and sudden events, supernova explosions, generate intense fluxes of neutrons, not allowing time for beta decay, thus synthesizing actinide elements and even heavier ones in the r-process (*r* for rapid). The critical information from the observations is the existence of elements heavier than the Pb group, implying that the r-process is a producer of the heaviest cosmic rays. The exact proportions of r- and s-processes for all the heavier-than-iron elements are still debated, but the presence of r is demonstrated.

OTHER USES OF COSMIC RAYS

In addition to being of interest in themselves, cosmic rays can be useful in other branches of science. Three examples lead us into lunar science, radiation dosimetry, and elementary particle physics.

Solar Cosmic Rays and Lunar Erosion

Solar cosmic rays (also called solar particles) include iron-group nuclei, and they come primarily from solar flares – magnetic storms on the surface of the sun. As a function of energy these particles have a steep distribution such that the tracks they form in solids in space are most abundant at surfaces, and rapidly decrease inward.

A gift of the Apollo program provided Price, Walker, and me with vital control data that has allowed erosion of lunar rocks to be measured. Figure 5-13 shows where, in November 19, 1969, Apollo 12 landed on the moon about 150 m from the Surveyor 3 spacecraft that had arrived 31 months before. The photo shows astronaut Charles Conrad, Jr., in the process of starting to remove the TV camera for return to earth. We had requested retrieval also of a large glass compartment cover that would be an excellent large-area track detector, equivalent in collection power to our lost flight over the Pacific, concentrated into a far smaller area. But the glass resisted removal. (Over 200,000 miles through space the radio signal relayed the eloquent voice of one of the astronauts, after a brief struggle with the cover, "Aw, forget it." Thus died that idea.)

Happily the camera head held a small but vital detector, a lead-glass optical neutral density filter, visible in the close up in Figure 5-14. For cosmic ray studies portions were allotted to Price at Berkeley, Walker at St. Louis, and H. R. Hart, G. M. Comstock, and me at General Electric. Each group (comfortingly) arrived at very nearly the same results for the cosmic ray spectrum (Crozaz and Walker, 1971; Fleischer *et al.*, 1971; Price *et al.*, 1971b). The right-hand side of Figure 5-15 shows the conclusion as to how rapidly cosmic ray tracks will accumulate with depth. Typical observations in lunar samples show significantly shallower drop offs; and these are interpreted, as shown on the left, as being due to gradual erosion of surfaces in space. Typical inferred erosion rates lie between 1 and 10 Å/yr (0.1 to 1 mm/M.y.), possibly from sputtering caused by the solar wind.

Radiobiological Effects of Cosmic Rays

At NASA's Manned Spacecraft Center in Houston, I was drowsing, along with other lunar sample investigators, through a briefing about lunar mission procedures and equipment for Apollo 11 when a simple statement brought me to sudden alertness. "The space helmets are made of Lexan polycarbonate." Our best-calibrated detector of cosmic rays! In the course of time this recognition led to our obtaining Apollo 8 and Apollo 12 helmets to measure the heavy particle doses to which the astronauts were exposed (Comstock *et al.*, 1971). In the experiments the helmets were inverted, partially filled with sodium hydroxide solution, etched for days, and then rinsed out.

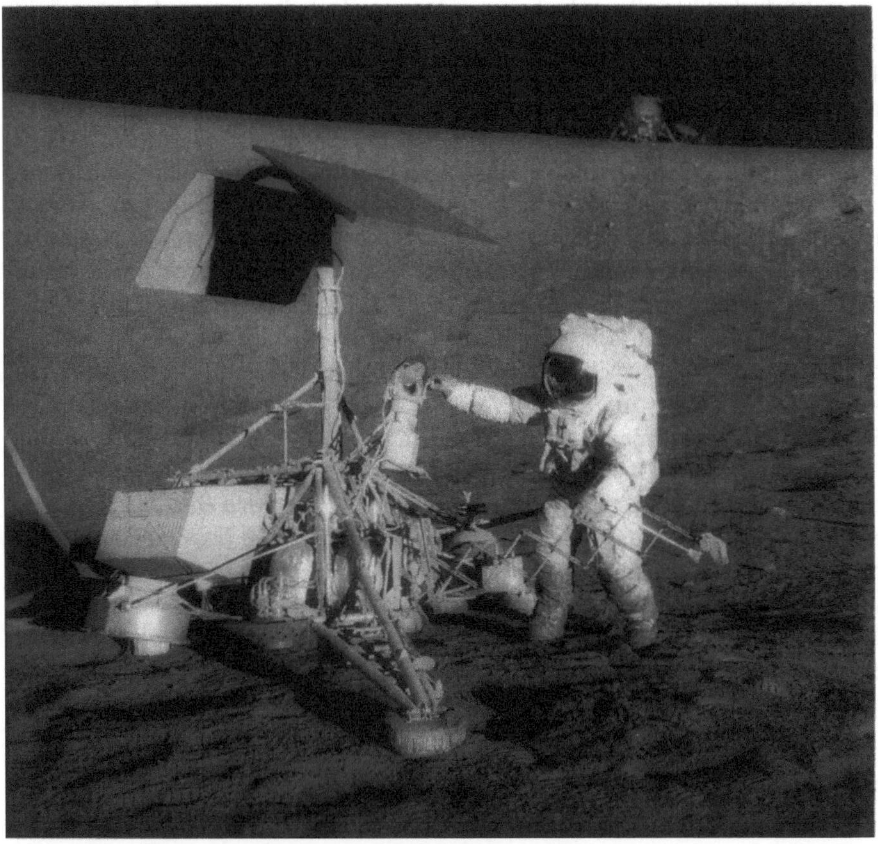

Figure 5-13. Apollo 12 astronaut Charles (Pete) Conrad, Jr., has strolled from the lunar module to the Surveyor 3 spacecraft and is about to remove the camera for return to earth (NASA photo by Charles Duke).

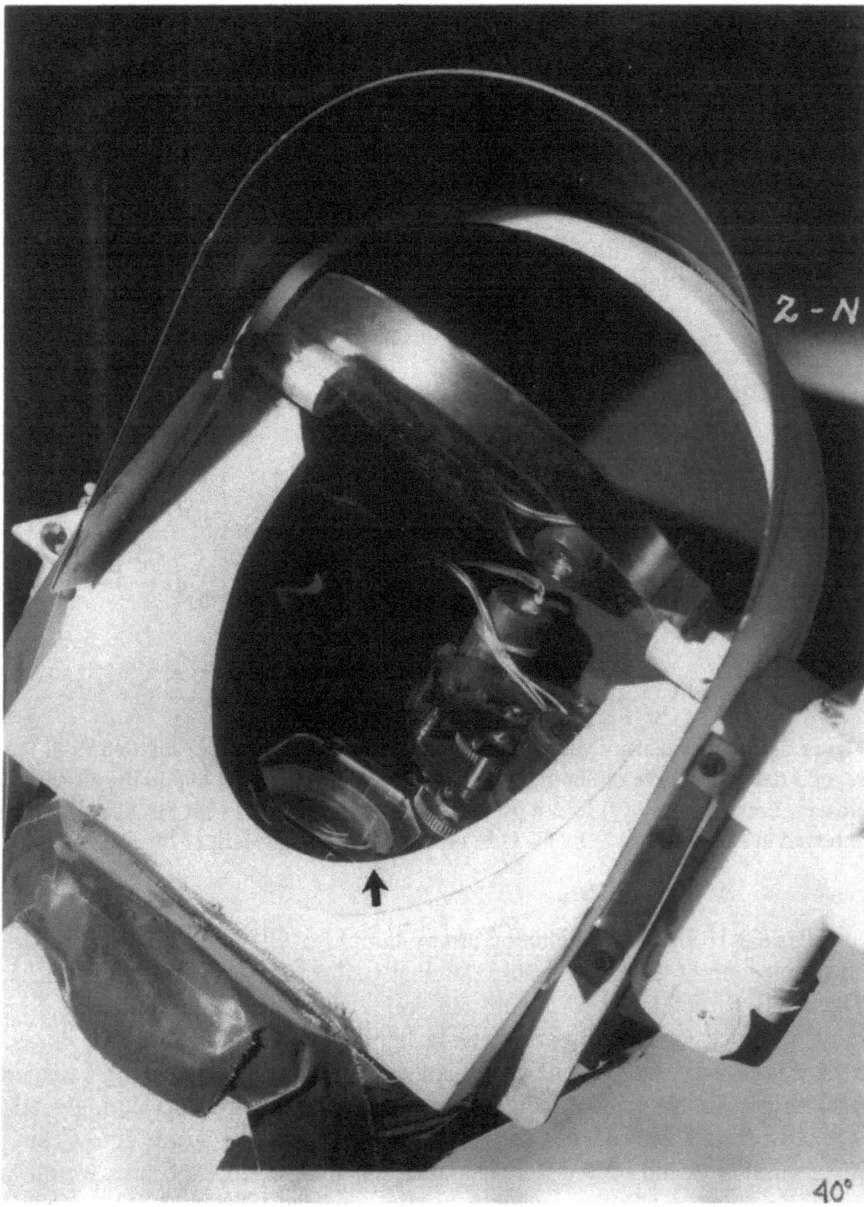

Figure 5-14. Portion of returned Surveyor 3 TV camera showing the flint glass filter located over the lens inside the shroud. Tracks in the glass provided a record of heavy cosmic rays for 2.55 years on the moon and allowed lunar erosion to be measured (NASA photo).

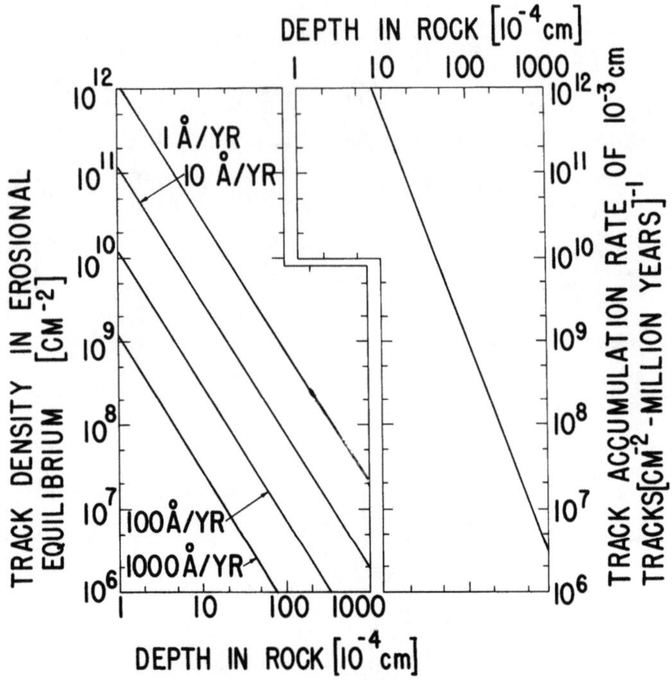

Figure 5-15. Track density versus depth as a function of erosion rate inferred from Surveyor 3 data. The curve on the right gives the track accumulation rate in the absence of erosion. Erosion rates of 0 to 2×10^{-8} cm yr^{-1} were consistent with the track densities observed in Apollo 11 rocks by the G.E. group (Fleischer and Hart, 1974).

Figure 5-16 shows the helmet worn by James Lovell on the Apollo 8 mission and a selection of the etched tracks. (Actually silicon-rubber casts of the tracks.) The heavy particles that we record are only a small part of the radiation dose, most of which is from dispersed, more lightly ionizing radiation that doesn't form tracks in Lexan. But one worry is that the intense damage along a nuclear track might disable critical motor-control cells of the astronauts during a mission. Our results showed that for typical lunar flights the danger is low, but a Mars mission would be a different matter, since expected exposures would be much longer, about two years in space. We obtained later results (Fleischer *et al.*, 1973) from the electrophoresis experiment on Apollo 14 and our own cosmic ray detector on Apollo 16. The fluxes of cosmic rays increased with time by a factor of 5 from Apollo 8 to Apollo 16 due to solar activity decreasing (as part of the normal 11-year cycle) and thus allowing more of the lower energy portion of galactic cosmic rays into the inner solar system.

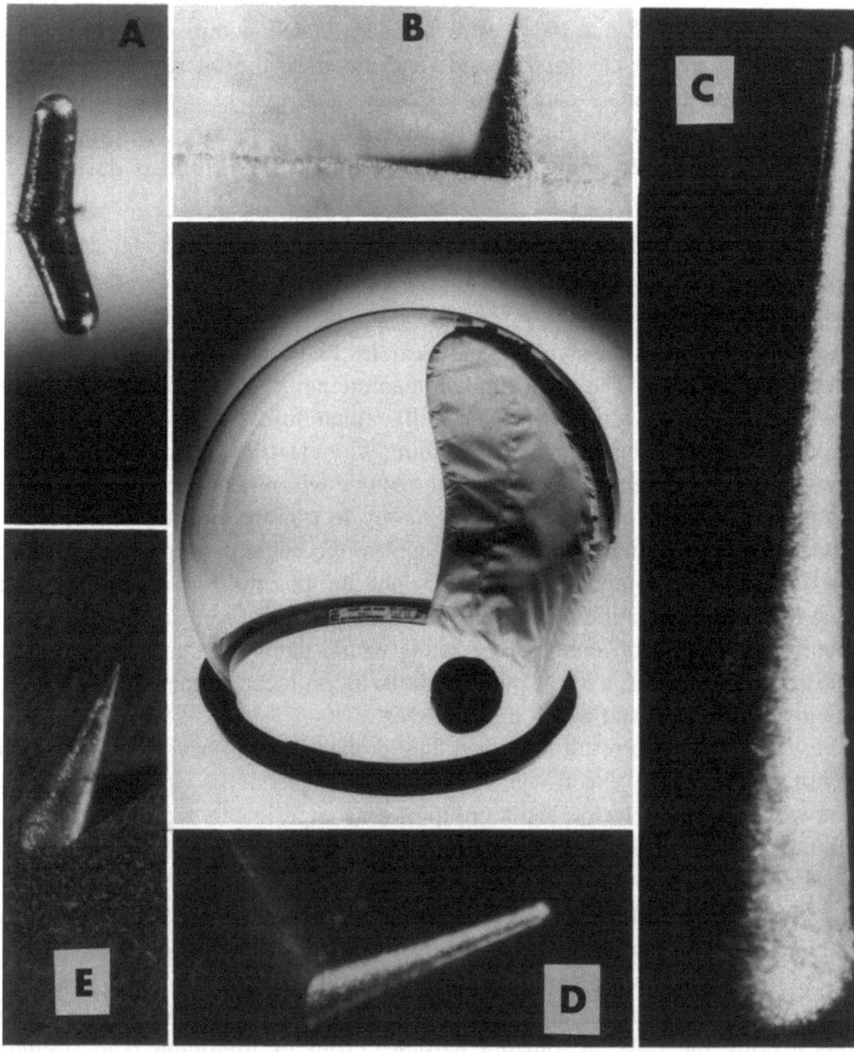

Figure 5-16. Cosmic ray tracks in Apollo helmets. The Apollo 8 helmet of astronaut Lovell, one of several used in a personnel dosimetry experiment, is surrounded by replicas of etched cosmic ray tracks. Tracks A, D, and E are from the helmets of Apollo 12 astronauts Conrad and Gordon, the other tracks from a control helmet exposed to primary cosmic rays. Track C is from a zinc ion. The lengths (in micrometers) of the track replicas are A, 350; B, 300; C, 700; D, 600; and E, 480. These tracks of heavy cosmic rays that penetrated the Apollo helmets provide a measure of the cytologically lethal component of the cosmic radiation. Tracks A, C, and D crossed the inside of the helmet (Fleischer *et al.*, 1970c).

The shapes of the tracks in Figure 5-16 are informative. The two sharp-pointed ones are what one normally expects, particles entering the helmet from outside. But the three with rounded tips are nuclei that came to rest *as they were leaving the helmet.* Since these were not accelerated within the helmet, they must have entered at the opposite side and traversed the interior. As scientists we have been studiously careful to reject the idea that in this work we were clairvoyant; nevertheless, Figure 5-16 shows that we have exact, quantitative evidence on what was going through the minds of the astronauts during their missions.

Magnetic Monopoles

Magnetic monopoles are hypothetical particles that are the magnetic analogues of electrical charges. The great English theoretician P. A. M. Dirac (1931) predicted them, first to symmetrize Maxwell's equations of electromagnetism and second to provide a basis for understanding why electrical charge is quantified. They have been sought in many ways in many environments, and such searches continue whenever someone identifies a way to a more thorough look. Had a lesser person than Dirac proposed monopoles, the attempts might well have tapered off, but their importance to theory and the possibility for an experimentalist of sending theoreticians back to the drawing board remains enticing. And unambiguous proof of monopoles existing would likely earn an immediate Nobel Prize. And there is a lot of pure fun in trying to locate something with many unknown properties and which may not exist.

Even if a positive result is of very low probability, the payoff would be immense. When an easy experiment is possible, most physicists would give it a try. And we at General Electric had a group of easy experiments we could try, since solid track detectors were particularly suitable for observing moving monopoles.

How do we know how to detect monopoles? We depend either on static properties of a pole (it is attracted and bound to magnetic materials, but can be extracted and accelerated by high magnetic fields) or on properties of moving poles. Just as a moving electrical charge creates a magnetic field, a moving magnetic charge would create an electrical field, which acts on nearby electrons, thus making the pole an ionizing particle. From its minimum pole strength, which Dirac inferred from a quantum relation between the electrical and magnetic charges, its track-forming ionization signature is equal to that of a fast nucleus of charge 68.5, 68.5 being half of the inverse of a scientifically well-known number, the fine-structure constant. In short, the etched cosmic ray track in Figure 5-11 is a calibration that tells us what we're looking for. But to avoid ambiguity we had best look where fast, heavy atomic nuclei in the cosmic rays are absent – for example, at or below the surface of the earth. A group of us that included Price looked in many places, including well-dated ferro-manganese crust from the ocean bottom and ancient mica (Figure 5-17). A summary of that work (Fleischer *et al.*, 1970d) and prior searches is given under the plaintive title "Magnetic Monopoles: Where Are They and Where Aren't They?" The answer was that we could eliminate a good many places by providing very low limits on monopole abundance.

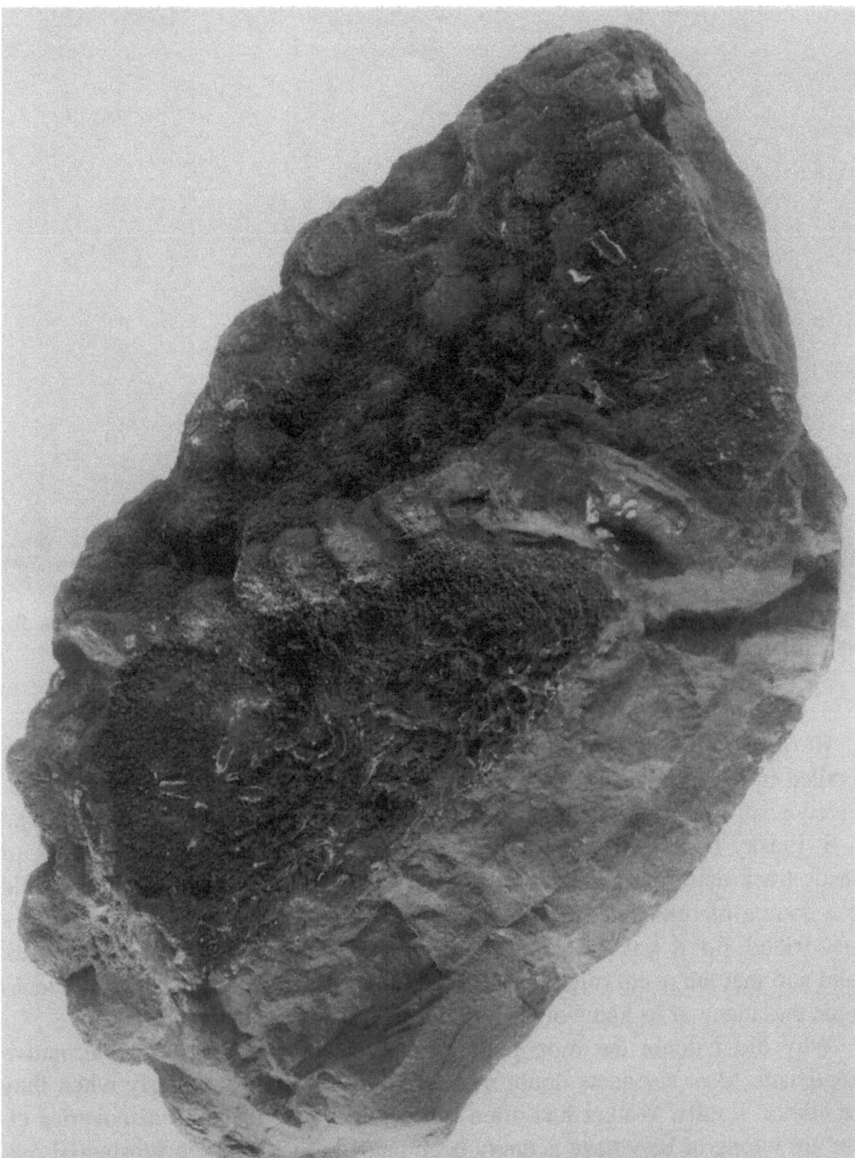

Figure 5-17. Potential monopole trapper and track recorder. a) Magnetic ferromanganese crust accumulated at the bottom of the North Atlantic Ocean during a 16-million year time span.

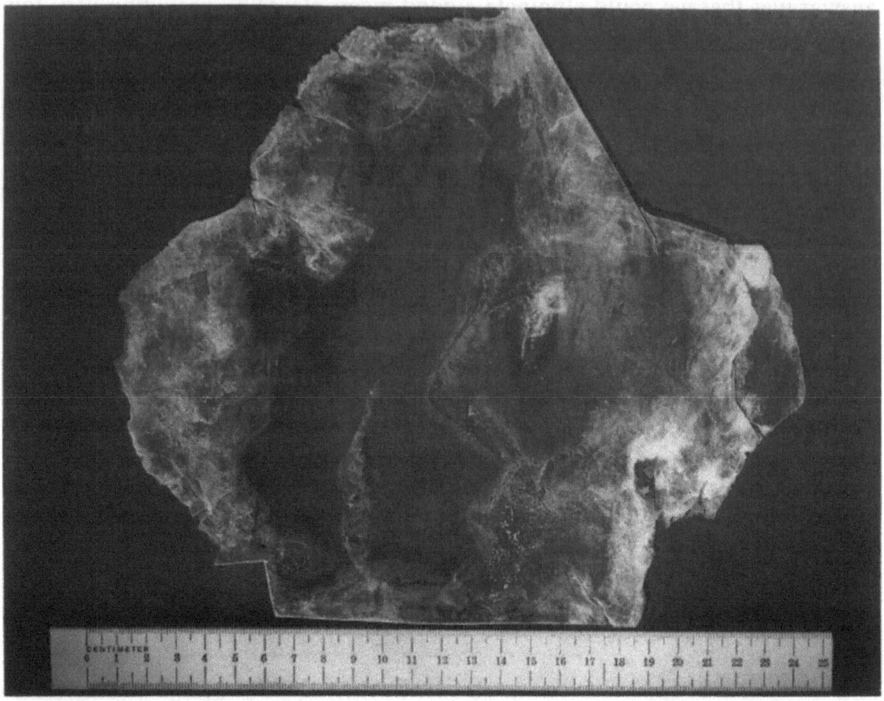

Figure 5-17. b) North Carolina Muscovite mica that has stored particle tracks for 248 million years (Fleischer *et al.*, 1970d).

However, my peace of mind on this subject was agonizingly and abruptly derailed in the summer of 1975. While on vacation at Martha's Vineyard, I received a call from Price telling me that he, with E. K. Shirk, W. Z. Osborne, and L. S. Pinsky had found the track of a monopole in a balloon-borne stack of plastic track detectors and other photographic films. The result, if true, should be a source of rejoicing that a long-sought particle is found, and found by a good friend. But it was agonizing in the likelihood that their conclusion was not valid and that the event might cast a shadow on the reliability of the track technique that many of us had worked so intensely to develop for many years.

Why did I doubt the monopole? The starting point I suspect was native skepticism. Most scientists doubt most startling results, particularly when they are others' results. Walker had often stated that most great new discoveries either are wrong or they have actually been made before. A more substantial reason for doubt was that the particle was seen on a cosmic ray flight where you have a great many cosmic rays as background to supply potential confusion.

And the sum of prior cosmic ray flights had a far greater gathering power (area times time), which required this event to be seen as the result of an extreme statistical fluctuation. And if you accept that extreme statistical fluctuations must occasionally occur, you might well imagine other such fluctuations that might cause conventional particles to sometimes mimic the behavior in detectors of a monopole.

After returning as soon as possible to the laboratory in Schenectady, we assembled an interested group to evaluate possible difficulties with the experiment, and to suggest what to look for in the as yet unseen data. The signature of a monopole is constant or decreasing ionization as it slows, whereas an ordinary nucleus increases in ionization. One way a nucleus might give a monopole-like signature would be to chip off some of its protons, reducing its ionization and converting it to a lighter particle, and this process might repeat.

On arriving in Berkeley and viewing the data shown in Figure 5-18, I felt that the data confirmed the possibility that such a nuclear interaction was involved. At 1.35 gm/cm^2 down into the stack there is a downward shift in etching rate followed by a gradual increase. The particle velocity is a critical factor; in the Price *et al.* (1975) work the absence of signal in a Cerenkov film was used to set an upper limit, but that type of film was new and therefore how stringent a limit it could set was not well founded. Are the data solely fluctuations about the solid line, or are there two or three slightly upward trends after abrupt drops? I recognized how the constant-value interpretation arose. The triangles in Figure 5-18 were the original data – the filled circles being the somewhat later completion of the available etched-track detectors. Given only the triangles, a horizontal line in Figure 5-18 gives a reasonable fit.

A number of reinterpretations were promptly supplied by L. Alvarez, by N. Durgaprasad and M. V. S. Rao, by P. H. Fowler, by M. W. Friedlander, and by Walker and myself. After reconsidering, Price shortly agreed that their evidence could not prove the event was a monople (Price, 1976), and after extensive, careful further analysis Price *et al.* (1978) were ready to deny that it was a monopole, but noted features that made it a highly unusual, difficult-to-explain event. Thus ended the monopole scare with propriety and honor – until 1981 when Cabrera (1982) recorded a monopole-like event, which his more extensive data also failed to confirm. A few years before his death I asked Dirac if the years of failure to find monopoles had dimmed his hopes. His response fit his reputation, "The reasons for believing in the magnetic monopole are as valid now as when I first suggested them."

COMMERCIALIZATION?

Notice the question mark in this heading. Cosmic rays have not been of direct commercial importance, at least to my former employer General Electric. But various uses of track detectors, including cosmic ray studies, have spawned business in supplying detector materials, particularly CR-39 (allyl diglycol carbonate), and LR115 and CN85 cellulose nitrates.

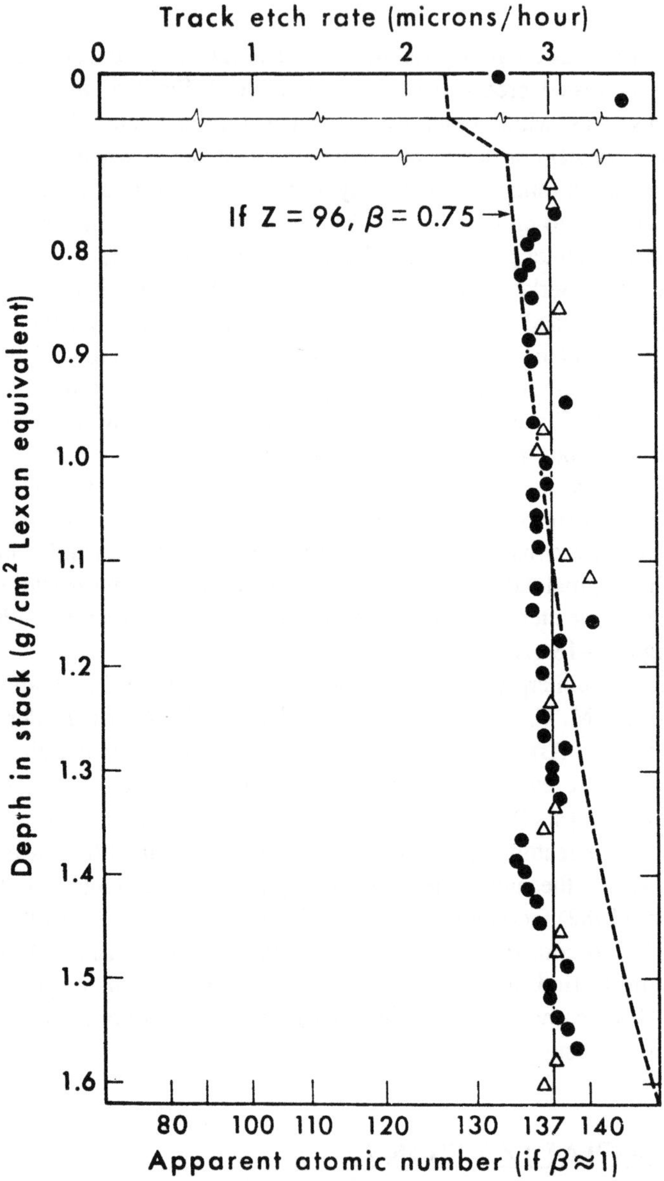

Figure 5-18. Original Lexan data for monopole candidate found on a cosmic ray flight. The upper two points are separated from the bottom two by 27 detector sheets. The solid line was the behavior expected of a charge 2 monopole; the dotted line that of a charge 96 nucleus with a speed of 0.75 c (c is the velocity of light) (Price *et al.*, 1975).

One might ask how the work described in this chapter was justified at General Electric. It was done partly on the basis that cosmic rays are evidence of energy-generation processes that are not well known and therefore might ultimately lead to practical energy-generation processes on earth. This view was received by management with amused tolerance, because payoff was regarded as equally remote in time as it is in space. The work really was encouraged as part of a longstanding tradition of contributions to science at General Electric, a tradition that has since been terminated, as is also true for other major industrial laboratories in the United States.

There is a continuing record, however, of scientific exploration and energy production being connected. Attempts at the use of fusion are derived from our understanding how the sun works. Oil is a major energy source because of the successes of geology and geophysics. So our justification was not frivolous, but rather too long range to satisfy present-day industrial managers.

Might Monopoles Be Useful?

Interestingly, practical uses of monopoles have been considered. The history of the search for the applications goes back to the first of the two purported discoveries of a monopole. Since General Electric is a commercial outfit, the statement was made there (and appropriately so) that if it was a real monopole and we had been trying to study them for some years and hadn't given attention to the possible applications, we would look a bit silly and even, perhaps, negligent. That view stimulated a brainstorming session at General Electric among C. P. Bean, M. D. Fiske, H. R. Hart, I. Giaever, and me. This group discussed various ideas, whose credibility can only be noted to exceed that given to the existence of monopoles. From our discussion emerged eight formal documents, patent disclosure letters, which have since been abandoned. Everyone contributed to the discussion, but the authorship of these patent disclosures turns out to be four of the five members. This group includes five fellows of the American Physical Society, three members of the American Academy of Arts and Sciences, three of the National Academy of Engineering, two members of the National Academy of Sciences, and one Nobel Prize winner. The only one that did not write a patent disclosure letter was in each of these groups, Ivar Giaever. It may be there really is something special about Nobel Prize winners; perhaps they have some judgment that the rest of us do not.

Ideas included high-density information storage, magnetically induced catalysis, high-energy accelerators and monopole generators, energy storage, and transportation – both by magnetic levitation and by sailing ships (tacking northward with a load of one sign of pole and southward after a change of poles). See Fleischer (1985) for more details, the reference appropriately being part of a symposium entitled “Applications of Particle Physics: Out on the Limb of Speculation.”

FURTHER READING ON COSMIC RAYS AND PARTICLE IDENTIFICATION

A popular-level article:

APT81 "Track-Recording Solids," S. P. Ahlen, P. B. Price, and G. Tarle, *Physics Today* **34**, 32–39 (1981) Sept.

More in depth:

FPW75 Chapters 3 and 5 of *Nuclear Tracks in Solids*, R. L. Fleischer, P. B. Price, and R. M. Walker, Univ. of California Press, Berkeley, 1975, pages 119 to 155 and 232 to 306.

Cosmic rays:

PF71 "Identification of Energetic Heavy Nuclei with Solid Dielectric Track Detectors: Applications to Astrophysical and Planetary Studies," P. B. Price and R. L. Fleischer, *Annr. Rev. Nucl. Sci.* **21**, 295–334 (1971).

REFERENCES

G. E. Blanford, M. W. Friedlander, J. Klarmann, R. M. Walker, J. P. Wefel, W. C. Wells, R. L. Fleischer, G. E. Nichols, and P. B. Price (1969), "Observation of Trans-Iron Nuclei in the Primary Cosmic Radiation," *Phys. Rev. Letters* **23**, 338–342.

E. M. Burbidge, G. R. Burbidge, W. A. Fowler, and F. Hoyle (1957), "Synthesis of the Elements in Stars," *Rev. Mod. Phys.* **29**, 547–650.

B. Cabrera (1982), "First Results from a Superconductive Detector for Moving Magnetic Monopoles," *Phys. Rev. Lett.* **48**, 1378–1381.

G. M. Comstock, R. L. Fleischer, W. R. Giard, H. R. Hart, Jr., G. E. Nichols, and P. B. Price (1971), "Cosmic Ray Tracks in Plastics: Apollo Helmet Dosimetry Experiments." *Science* **172**, 154–157.

G. Crozaz and R. M. Walker (1971), "Solar Particle Tracks in Glass from the Surveyor 3 Spacecraft," *Science* **171**, 1237–1239.

P. A. M. Dirac (1931), "Quantized Singularities in the Electromagnetic Field," Proc. R. Soc. Lond. **A133**, 60–72.

R. L. Fleischer (1985), "Magnetic Monopoles," in R. A. Carrigan, ed., *Fermilab Industrial Affiliates Roundatble: Applications of Particle Physics: Out on the Limb of Speculation.* Batavia, Fermi National Accelerator Laboratory, May, pp 65–77.

R. L. Fleischer and H. R. Hart (1974), "Particle Track Record of Apollo 16 Rocks from Plum Crater," *J. Geophys. Res.* **79**, 766–768.

R. L. Fleischer, P. B. Price, and R. M. Walker (1965), "Solid State Track Detectors: Applications to Nuclear Science and Geophysics," *Ann. Rev. Nuc. Sci.* **15**, 1–28.

R. L. Fleischer, P. B. Price, R. M. Walker, and M. Maurette (1967a), "Origins of Fossil Charged Particle Tracks in Meteorites," *J. Geophys. Res.* **72**, 333–353.

R. L. Fleischer, P. B. Price, R. M. Walker, M. Maurette, and G. Morgan (1967b), "Tracks of Heavy Primary Cosmic Rays in Meteorites," *J. Geophys. Res.* 72, 355–366.

R. L. Fleischer, P. B. Price, R. M. Walker, R. C. Filz, K. Fukui, M. W. Friedlander, E. Holeman, R. S. Rajan, and A. S. Tamhane (1967c), "Observations of Cosmic Rays in Plastics," *Science* **155**, 187–189.

R. L. Fleischer, E. L. Haines, H. R. Hart, Jr., R. T. Woods, and G. M. Comstock (1970a), "The Particle Track Record of the Sea of Tranquility," *Geochim. Cosmochim. Acta* **3**, Suppl. I, 2103–2120.

R. L. Fleischer, E. Lifshin, P. B. Price, R. T. Woods, R. W. Carter, and E. L. Fireman (1970b), "Schenectady Meteorite," *Icarus* **12**, 402–406.

R. L. Fleischer, H. R. Hart, and W. R. Giard (1970c), "Particle Track Identification: Application of a New Technique to Apollo Helmets," *Science* **170**, 1189–1191.

R. L. Fleischer, H. R. Hart, I. S. Jacobs, P. B. Price, W. M. Schwarz, and R. T. Woods (1970d), "Magnetic Monopoles: Where Are They and Where Aren't They?" *J. Appl. Phys.* **41**, 958–965.

R. L. Fleischer, H. R. Hart, and G. M. Comstock (1971), "Very Heavy Solar Cosmic Rays: Energy Spectrum and Implications for Lunar Erosion," Science **171**, 1240–1242.

R. L. Fleischer, H. R. Hart, G. M. Comstock, M. Carter, A. Renshaw, and A. Hardy (1973), "Apollo 14 and 16 Heavy Particle Dosimetry Experiments," *Science* **181**, 436–438.

P. H. Fowler, R. A. Adams, V. G. Cowen, and J. M. Kidd (1967), "The Charge Spectrum of Very Heavy Cosmic Ray Nuclei," Proc. Roy. Soc. Lond. **A301**, 39–45.

P. H. Fowler, V. M. Clapham, V. G. Cowen, J. M. Kidd, and R. T. Moses (1970), "The Charge Spectrum of Very Heavy Cosmic Ray Nuclei," Proc. Roy. Soc. Lond. **A318**, 1–43.

P. S. Freier, E. J. Lofgren, E. P. Ney, F. Oppenheimer, H. L. Bradt, and B. Peters (1948), "Evidence for Heavy Nuclei in the Primary Cosmic Radition, " *Phys. Rev.* **74**, 213–217.

M. Maurette, P. Pellas, and R. M. Walker (1964), "Cosmic Ray Induced Particle Tracks in a Meteorite," *Nature* **204**, 821–823.

D. O'Sullivan, P. B. Price, E. K. Shirk, P. H. Fowler, J. M. Kidd, E. J. Kobetich, and R. Thorne (1971), "High Resolution Measurements of Slowing Cosmic Rays from Fe to U," *Phys. Rev. Lett.* **26**, 463–466.

P. B. Price (1976), "Status of the Evidence for a Magnetic Monopole", in A. Perlmutter, ed., New Pathways in High-Engery Physics, NY, Plenum Publishing Corporation, Vol. **1**, pp 167–214.

P. B. Price and R. L. Fleischer (1970), "Particle Identification by Dielectric Track Detectors," *Radiat. Effects* **2**, 291–298.

P. B. Price, R. L. Fleischer, D. D. Peterson, C. O'Ceallaigh, D. O'Sullivan, and A. Thompson (1967), "Identification of Isotopes of Energetic Particles with Dielectric Track Detectors," *Phys. Rev.* **164**, 1618–1620.

P. B. Price, P. H. Fowler, J. M. Kidd, E. J. Kobetich, R. L. Fleischer, and G. E. Nichols (1971a), "Study of the Charge Spectrum of Extremely Heavy Cosmic Rays Using Combined Plastic Detectors and Nuclear Emulsions," *Phys. Rev.* **D3**, 815–823.

P. B. Price, I. D. Hutcheon, R. Cowsik, and D. J. Barber (1971b), "Enhanced Emission of Fe Nuclei in Solar Flares," *Phys. Rev. Lett.* **26**, 916–919.

P. B. Price, E. K. Shirk, W. Z. Osborne, and L. S. Pinsky (1975), "Evidence for Detection of a Moving Magnetic Monopole." *Phys. Rev. Lett.* **35**, 487–490.

P. B. Price, E. K. Shirk, W. Z. Osborne, and L. S. Pinsky (1978), "Further Measurements and Reassessment of the Magnetic-Monopole Candidate," *Phys. Rev.* **D18**, 1382–1421.

Proceedings of the Apollo 11 Lunar Science Conference (1970), A. A. Levinson, ed., *Supplement I, Geochim. Cosmochim. Acta*, Vol 3, "Physical Properties."

Chapter 6
DIVERGENT TRACK USES

There may be something interesting in this container.
Pandora, *source unidentified*

"You know something that you wouldn't talk about if you realized its significance," Price, Walker, and I were told one day at General Electric. But unfortunately we didn't have security clearances, and so we couldn't be told what we wouldn't talk about if we knew more about it. The result was that each of us was presented with apparently infinitely long clearance forms to make out: "Where did you live and for how long?" At all times in the past. And such.

In due course we received clearances, and that was the end. We heard nothing more. Events like that naturally serve to excite ones curiosity; and after some speculation, our bet was that use of track etching to detect and learn about nuclear weapons tests was at issue, since we had hints of such work by G.E.'s Vallecitos Atomic Laboratory, and also a somewhat related project at the Neutron Devices Department in St. Petersburg, *Florida.* That sort of rather simple use of tracks, to be described shortly, was one commercial application by General Electric, but we will never know how much it profited the company nor how much human effort was involved. But the techniques that were used are a part of the main theme of this chapter.

INDUCED TRACKS CARRY MANY MESSAGES

In Chapter 4, I described how the uranium content of a sample to be dated is found by using slow neutrons from a nuclear reactor to induce fission of uranium, thus creating new tracks. If we think about that process briefly, we realize that the track density created ρ_i depends not only on the uranium content c(u), but also on the dose of neutrons ϕ and the cross section for fission σ – the effective area of the nucleus of ^{235}U over which a slow neutron causes the nucleus to split. The simple proportionality $\rho_i \sim c(u)\ \phi\ \sigma$ tells us of three fields of use of tracks. Where we specify ϕ and know σ, the track density tells us c(u). So we can do chemical analysis. Where σ and c(u) are known, the dose can be measured. And where ϕ and c(u) are known, we can do nuclear physics in the form of measurements of nuclear cross sections. There are many applications that use one of these three possibilities.

NUCLEAR PHYSICS

The variety of nuclear physics uses that have been pursued is indicated by Table 6-1. For each item substantial numbers of papers are cited in FPW75. Items 1–6 are all similar to the example given of neutron-induced fission of uranium in that a cross section can be measured for a given reaction from the irradiation dose and the number of target nuclei that are present. The projectiles differ both in energy and character. They range from neutrons, to protons, to heavier charged particles, photons, and electrons. The special virtues of solid track detectors are 1) the ability to record the events of interest and 2) the insensitivity that prevents the observer from seeing the projectiles themselves, since, if they were imaged, the detector would be overwhelmed by their abundance, and the interesting, occasional events would be obscured.

Table 6-1. Areas of Nuclear Physics Where Track Etching is Used

1. Neutron-induced fission
2. Charged-particle-induced fission
3. Photofission and electrofission
4. Ternary fission
5. High-energy reactions
6. Low-energy reactions
7. Spontaneous fission
8. Spontaneously fissioning isomers
9. Compound nucleus lifetimes by blocking
10. Searches for superheavy elements

Ternary Fission

As an example of useful insensitivity, Figure 6-1 shows tracks induced in a crystal of thorite ($ThSiO_4$) that was irradiated by ^{40}Ar ions. Note that a typical event would easily fit into an area of 20-μm diameter in Figure 6-1. The exposure to argon ions is approximately 30,000 ^{40}Ar ions entering the crystal in each such area. Only the occasional interaction is seen where Ar merges with Th and the resultant excited, moving nucleus blows apart, usually into two parts, but occasionally (3% of the time) into three parts, a new behavior for which the track technique made possible direct visualization of the events. The binary fissions in the figure have V-shapes because the momentum imparted by the argon gives the fissioning nucleus a velocity, so that it splits in flight.

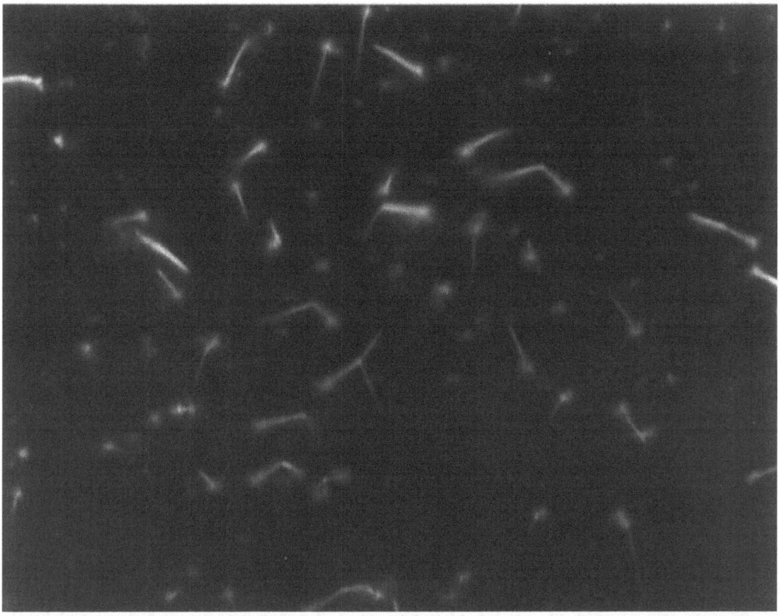

Figure 6-1. Etched charged-particle tracks in thorite that was irradiated with 400-MeV Ar ions. The two-pronged events are binary fissions, the three-pronged event is ternary fission. Typical prongs are 5–10 μm long. The Ar beam incident at 9° to the horizontal does not produce etchable Ar tracks. If it produced tracks, the whole field of view would be lighted up with parallel tracks, and the events of interest would be obscured (Fleischer *et al.*, 1966).

That high sensitivity may be undesirable and may seem surprising. But consider your problem if you were given the phone directory of a large city and told to look up 100 specific people, but found that names were given at random. A small directory with only the names you want would look like a luxury.

In principle the fission process might be a prompt process, in which the projectile knocks the nucleus and goes on as the nucleus comes apart. The alternative is what actually happened here, where the Ar merged with the Th nucleus forming temporarily an excited nucleus of element 108, which then vibrates itself apart. In this process, called compound nucleus formation, the collision is in general off-center of the target, thus causing the compound nucleus to tumble around an axis that is perpendicular to the beam direction. The effect of this rotation is to create a forward-backward preference to the fission particles, as experimentally demonstrated in Figure 6-2 (Fleischer *et al.*, 1966).

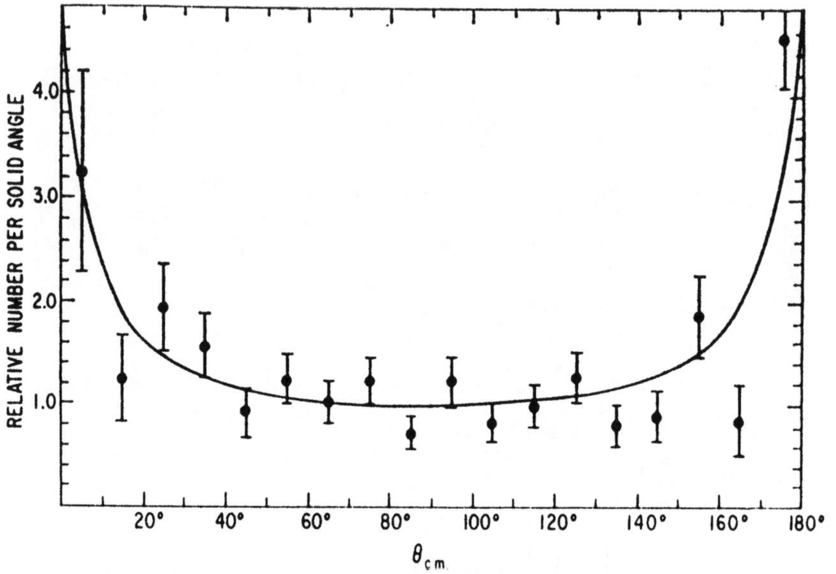

Figure 6-2. The angular distribution relative to the beam direction of the orientation of prongs of ternary events in center-of-mass co-ordinates shows the forward-backward preference of a tumbling nucleus (Fleischer *et al.*, 1966).

Track detectors have been used to measure minute cross sections, down to 10^{-35} cm^2 (Burnett *et al.*, 1964), and to visualize other interesting nuclear processes (Figure 6-3).

V-Tracks and Cosmic Rays

For extraterrestrial objects it is often of interest to know how long they have been in space in their present form, a period that is measured by how long cosmic ray effects have accumulated. In the case of some tektites there is one rather direct way that is suggested by their shapes.

Figure 6-4 illustrates the situation. Round tektites that are ablated on one side as they enter the atmosphere are reshaped, with a prominent feature being the flange that circles the ablated face. The figure shows both a naturally shaped tektite and one that was ablated in the laboratory to reproduce formation conditions with remarkable fidelity. It was this elegant work by Chapman and Larson (1963) that established that the shaping did occur during entry to the earth's atmosphere.

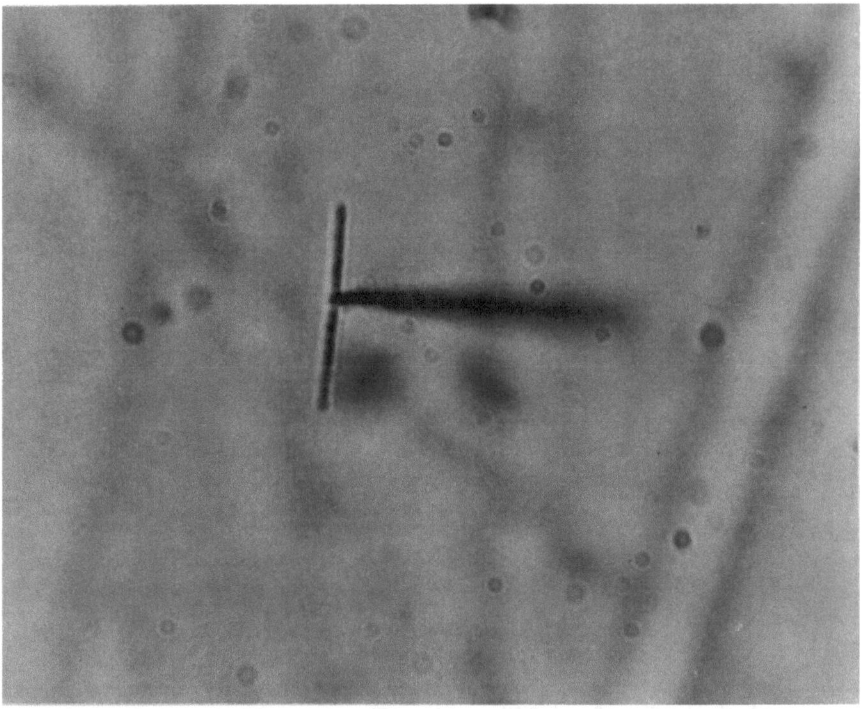

Figure 6-3. "Hammer" track from a light particle that disintegrates as it comes to rest in a cellulose nitrate detector. The particle is a light ternary fragment from fission of ^{252}Cf (most likely ^{8}He). The fragment decays to ^{8}Be, which in turn promptly decays into two ^{4}He nuclei (courtesy of P. B. Price).

An obvious test of how long tektites spent in space is to measure the age of the core and of the flange of tektites – since the first is the birth date and the second the time of arrival at the earth. We did this for four flanged tektites and found no difference in age (Fleischer and Price, 1964). This result does not mean zero time in space, merely that the maximum time in space that could have occurred was about 50,000 years, the resolution of the method.

A far tighter restriction of the space exposure comes from seeking V-tracks. In addition to the compound nucleus reactions described, V-shaped tracks can come about in a prompt reaction, such as might be caused by a relativistic cosmic ray proton. Here the momentum that makes the track V-shaped, rather than straight, is imparted by a few neutrons and protons that are rather promptly ejected from a heavy nucleus, which consequently recoils and fissions in flight. Protons are the most abundant species in the cosmic rays; and, by this process, they can cause distinctive tracks that are a measure of how long a sample was exposed in space.

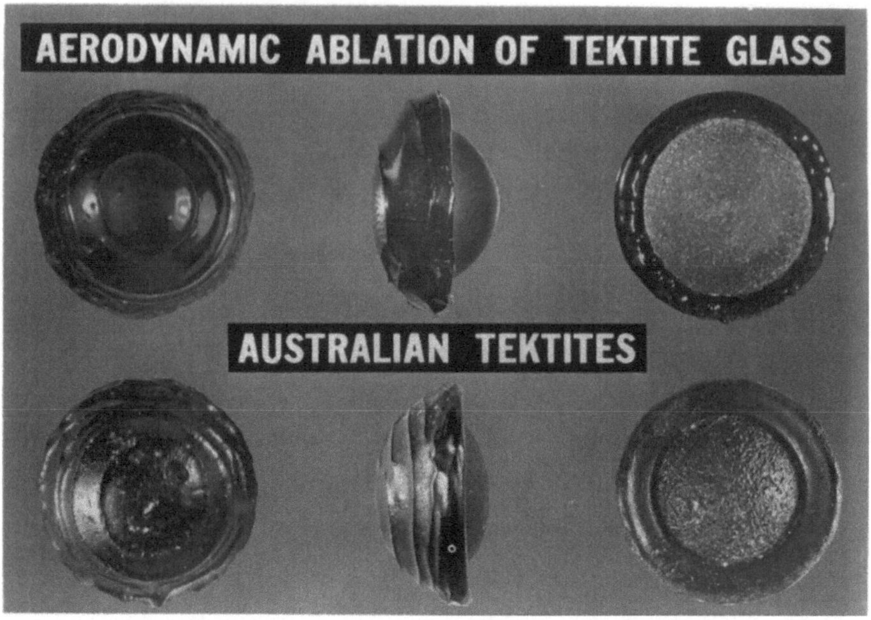

Figure 6-4. Flanged tektites were produced during atmospheric entry of solid round tektites. Below: Natural tektites. Above: Flanged shapes produced in an arcjet to simulate the atmospheric shaping of an infalling object (Chapman and Larson, 1963; photo courtesy of Marguerite Chapman).

Figure 6-5 shows how a tektite is calibrated by irradiation with a known dose of fast protons from an accelerator. The V's are less obvious in a glass object than they are in Figure 6-1, but the altered shapes are nevertheless distinctive; i.e. compare these shapes with those in Figure 4-1. We found from such experiments that 8 tracks/cm^2 would result from 10^{12} protons/cm^2 (Fleischer *et al.*, 1965a). However, we found no V-tracks in surveying 2.6 cm^2 of unirradiated tektite. From the known flux of cosmic ray protons this result sets a limit that the typical tektite spent less than 300 years in space. Tektites thus have remarkably different records from the meteorites, which have spent millions of years in space as small objects. The findings confirm the nearby origin of tektites, but does not tell us whether they started at the moon or at the earth – since both travel times are much less than 300 years.

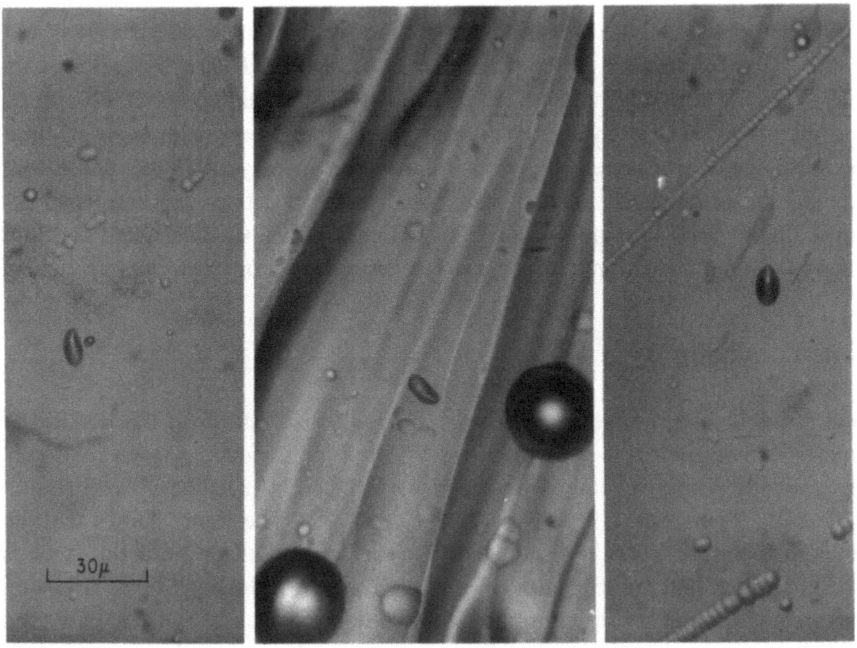

Figure 6-5. V-shaped etch pits in a tektite that was irradiated with 3×10^9 eV protons. This is a tektite from Indochina etched 85 sec in 48 HF. The suface was normal to the proton beam. The etched V-tracks are bent ellipses having in these three cases lengths of about 10 μm (Fleischer *et al.*, 1965a).

A second case where V-tracks can be related to exposure to space is the Surveyor 3 glass filter discussed in Chapter 5. At depth in the glass the solar particles had been stopped, and only occasional fission tracks remained – ones that looked very much like those in Figure 6-5 (Fleischer *et al.*, 1971). That particular glass is unusual in being rich in a heavy element, lead, which can be fissioned by fast galactic protons and alpha particles.

Thus, we have seen manmade V's in natural materials and natural V's in a manmade material. As yet I am aware of no clear cases of naturally induced V's in natural materials.

CHEMISTRY

Uranium can be analyzed by inducing fission by thermal neutrons (Price and Walker, 1963). Similarly, other elements can be made to fission by various particles: ^{239}Pu by both thermal and fast neutrons, ^{232}Th by fast neutrons, Pb and Bi by alpha particles or high-energy protons. Hence the presence of each of these elements can be measured and their distribution displayed. If the phase that contains the elements of interest will record the tracks, they can be etched directly. Otherwise a low-uranium detector is pressed against a flat surface of the material during irradiation to receive fission fragments from the material.

Superconductivity

Figure 6-6 is an example in which the uranium distribution in the surface of superconducting wires is evident (FAFPW72). Superconducting high-field current-carrying properties in compounds are enhanced by irradiation with fission fragments (Bean *et al.*, 1966; Fleischer *et al.*, 1989). The figure shows that the uranium is present where it's wanted, in the surface regions of the wire, which is where the intermetallic superconducting phase Nb_3Sn is located. Now, knowing how much uranium is present in the Nb_3Sn, a dose of neutrons will be delivered that produces the desired level of fission fragment damage in the superconductor.

Radionuclides and Biology

A radiobiological use of induced tracks is illustrated by Figure 6-7, in which neutrons have induced fission of ^{239}Pu that had been injected into a dog (Jee *et al.*, 1972). Lexan polycarbonate was placed against a section of trabecular (porous) bone and neutron irradiated to produce fission tracks – the dark lines that show that the plutonium coats the bone surfaces. Since surfaces are where new bone cells form, this location is a specially destructive one.

Boron

The bone in Figure 6-7, imaged by the method of Jee *et al.* (1972), is displayed and distinguished from the adjacent tissue by many small dots that provide chemical mapping of another element, B (Bean *et al.*, 1966). The boron isotope ^{10}B when irradiated with thermalized neutrons undergoes a break-up that is much like a miniature fission into an alpha particle and a ^{7}Li, a process written as $^{10}B(n, \alpha)^{7}Li$. Both ^{7}Li and the alpha can produce tracks in Lexan, and it is the many short tracks from boron in bone that appear in the figure and darken the bone area.

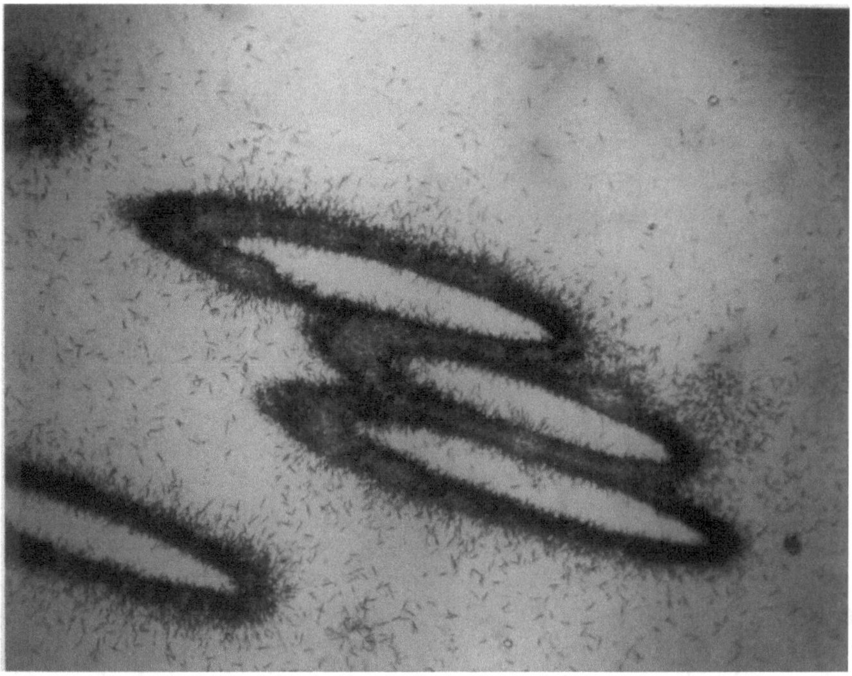

Figure 6-6. Micrograph showing uranium distribution in taper-sectioned 0.005-cm superconducting Nb_3Sn wires. Uranium-enriched tin was diffused into a thick outer section of the wire. Tracks from induced fission of uranium clearly delinate the thickness of the uranium-rich layer (FAFPW72).

Lithium itself can be mapped by the $^{6}Li(n, \alpha)^{3}H$ reaction, where both the alpha particle and the tritium nucleus can be recorded in cellulose nitrate or allyl diglycol carbonate detector.

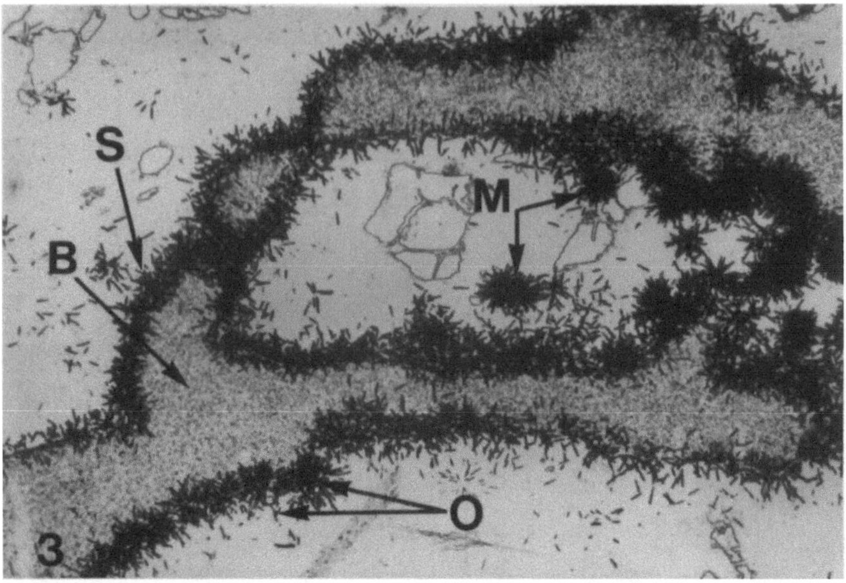

Figure 6-7. This neutron-induced radiograph shows the location of plutonium in the lumbar vertebra of a dog. The dog was given 2.7 μCi of ^{239}Pu/kg and killed 28 days later. A Lexan polycarbonate detector was pressed against a section from the lumbar vertebra, neutron irradiated, and then etched to produce this image, which shows the plutonium concentrated on bone surfaces S and in star-producing aggregates within the marrow M. Typical tracks are ~10 μm in length. Many minute alpha tracks from boron produce the dark image of the bone B (courtesy of W. S. S. Jee).

Chemical Analysis of Liquids

U, Pu, or B in liquids can be measured simply by evaporating a drop of known volume on a detector and then neutron irradiating to induce fission or an (n, α) reaction (Fleischer and Lovett, 1968), with results like those shown in Figure 6-8. Using clean detectors and water, and high neutron exposures, groups of as few as fifty, ^{239}Pu atoms have been measured (Fleischer and Raabe, 1977).

Where the liquid has other major constituents beyond the one of interest, it may be necessary to perform chemical separations before drop deposition. Using separations and great care in cleanliness, a group at Brookhaven National Laboratory (Moorthy *et al.*, 1988) found they could measure as few as 4×10^6 atoms of ^{239}Pu in urine. Such sensitivity is important in assessing uptake and elimination of plutonium in people who have been exposed to it – for example, the Marshall Islanders who encountered fall-out from nuclear weapons tests in the Pacific Ocean.

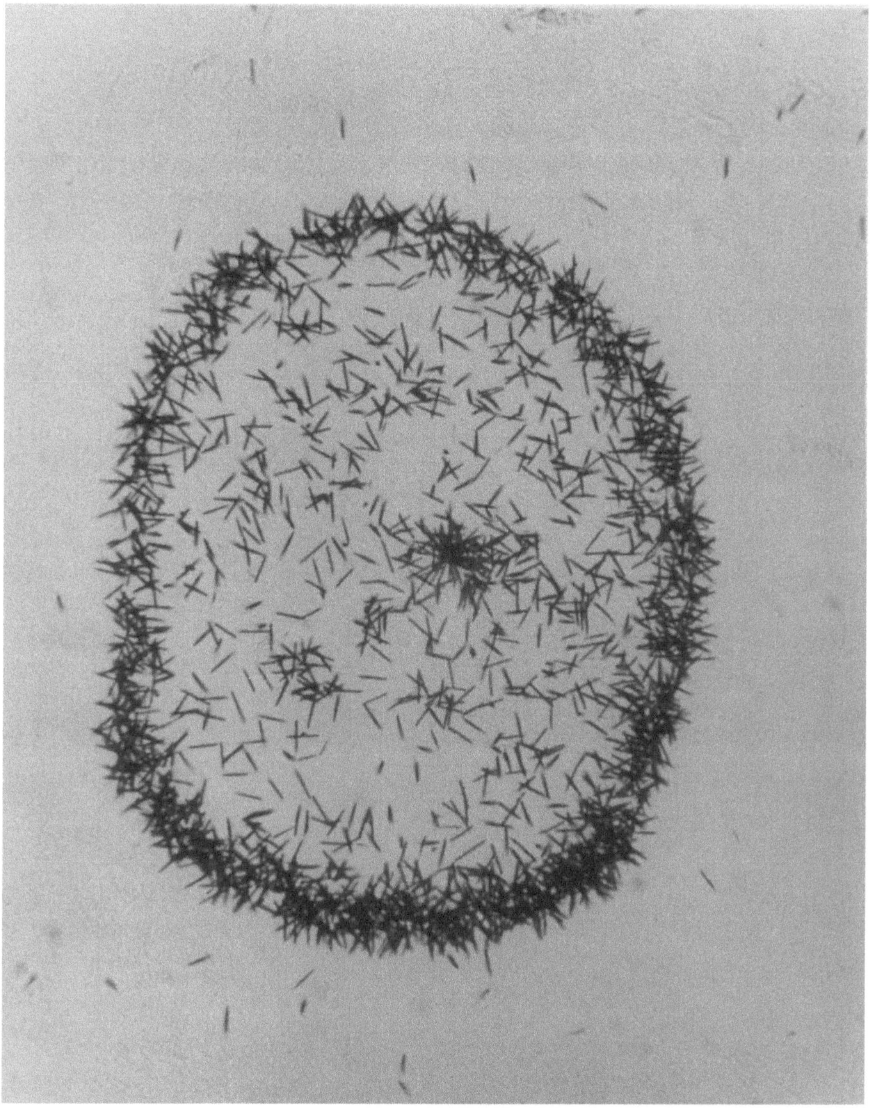

Figure 6-8. Uranium in water can be measured by allowing a droplet of known volume to evaporate on a track detector and then revealing the uranium by using neutrons to induce fission. In one view (this page) the edge of the droplet is outlined. In the other (following page) most of the uranium is in particles. Typical tracks are 18 μm in length (Fleischer and Delany, 1976).

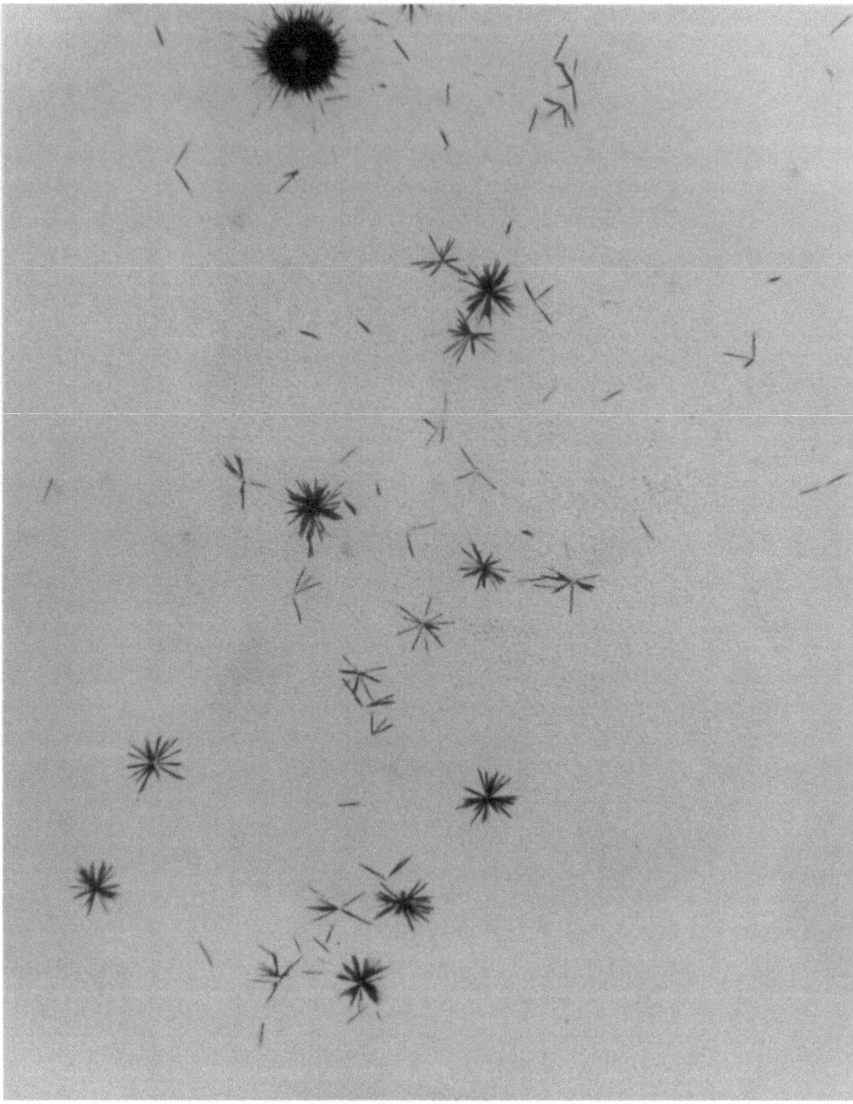

Figure 6-8. (continued)

Geochemistry

Rock samples can be analyzed in the same way as bone and tissue, as pictured in Figure 6-9 (Kleeman and Lovering, 1967). On the left is a rock section and on the right a track radiograph that is analogous to the one in Figure 6-7. The fission tracks show that virtually all of the uranium is in the white phase, and the shading shows that boron is distributed through both that phase and the adjacent slightly darker one.

Identification of Transuranic Radionuclides

At the onset of this chapter is was noted that aerosol particles have been used to learn about atmospheric tests of nuclear weapons. How can tracks identify what radionuclides are present in atmospheric particles?

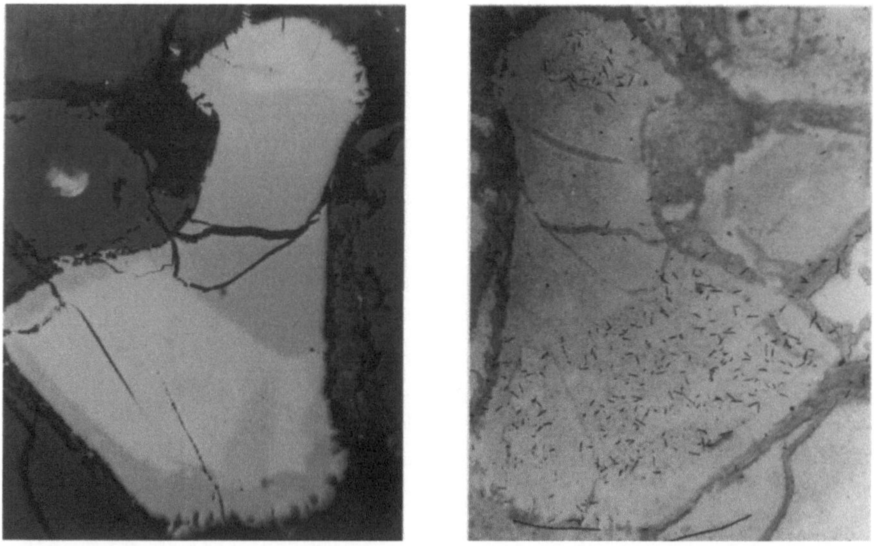

Figure 6-9. Optical photograph (left) and Lexan print (right) of a terrestrial rock. Variations in the unresolvably high density of induced alpha tracks in different minerals and in grain boundaries provide an easily recognizable image of the rock. The fission track distribution shows that the light phase, rutile (TiO_2), contains 1.1 parts per million of uranium (Kleeman and Lovering, 1967). The longest tracks are 18 μm in length.

One answer depends on the fact that many heavy elements both spontaneously alpha decay and can be induced to fission. That pair of properties can be diagnostic (Center and Ruddy, 1976). In Figure 6-10 at the left are alpha tracks in a piece of cellulose nitrate on which a particle was left for a specified time to record natural alpha decay. On the right are induced fission tracks from irradiation of the particle with a specified neutron dose. Since the number of atoms is fixed when the experiments are done on the same particle, the ratio of the tracks depends only on the ratio of two nuclear properties – the rate of alpha decay λ and the cross section for fission σ. And discrimination of pure radionuclides is easy since this ratio λ/σ ranges widely: Relative to natural uranium, U that is 95% enriched in ^{235}U has a ratio of 0.046, ^{238}Pu 1.3×10^6, ^{239}Pu 104, and ^{241}Am 1.4×10^7. Mixtures of these radionuclides will have intermediate values.

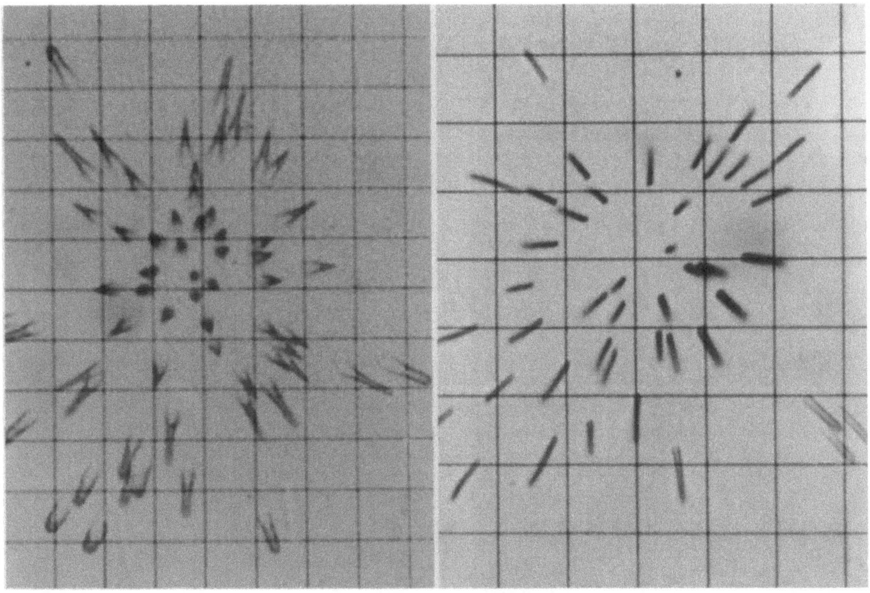

Figure 6-10. Tracks from a small particle that contains plutonium. Left: Tracks from spontaneous alpha decay. Right: Neutron-induced fission tracks. The ratio of the track densities is proportional to the ratio of the alpha decay constant to the fission cross section – a nuclear property that identifies the radioactive species (Center and Ruddy, 1976).

Miscellaneous Technological Uses

Boron is of industrial importance in metallurgical and ceramic materials. In diamonds it adds a desirable blue tint; in stainless steel it keeps carbon out of grain boundaries where embrittlement might otherwise occur; in silicon steel it stabilizes grain boundaries that help keep magnetic domains easily mobile for efficient alternating current transformers. In each of these areas boron mapping using tracks has supplied quantitative microstructural information that has aided industrial processes (Chrenko, 1971; Scanlon, 1975; Armijo and Rosenbaum, 1967).

DOSIMETRY

Measuring doses of particles is important to nuclear technology and human biological safety. Parts of Chapter 3 covered major topics in this area, examples being the facts that radon dosimetry in homes and mines has direct negative consequences for human health, and identifying radon concentrations in the ground is a method for locating the raw material for nuclear fuel.

Neutron Dosimetry

Measuring neutrons is a second dosimetry area of major track activity – again used both for radiobiology and for nuclear technology. There are two approaches, either observing direct neutron effects in the detector or counting induced reactions whose emitted particles are detected.

For personnel dosimetry, direct effects are most commonly used. Here one selects a sensitive plastic detector, such as cellulose nitrate, allyl diglycol carbonate, or bisphenol polycarbonate in which a fast neutron may elastically scatter off a C, N, or O nucleus and transfer sufficient momentum that the struck nucleus can leave a long enough track to be observed. The more numerous such collisions are, the more tracks occur; and the higher the energy of the neutron is, the longer the average track will be. So both fluence and energy information are created – two factors that contribute to biological effects.

Commercialization of Neutron Dosimetry

Practical application here started at R. S. Landauer, which produces Neutrak dosimeters; they use the direct technique just described.

There are inherent uncertainties with this method because the tracks are mostly short, so that obtaining reproducible results depends on carefully standardizing readout conditions – including etching time and temperature, etchant composition, and optical illumination and magnification. At higher magnification, for example, more tracks will be recognizable; but scanning is more laborious. And further, if the flux is in the form of a beam, the track count depends on the angle between the beam and the plane of the detector. In one assessment of the reproducibility of the technique, Fleischer *et al.* (1984) found a ± 30% to 40% uncertainty in the biological doses. That doesn't sound like outstanding accuracy, but it is the best available for fast neutrons.

Nuclear Reactions

Induced reactions, as shown earlier in this chapter, can give crisp, unambiguous signals (Walker *et al.*, 1963). Starting with this idea, Prevo *et al.* (1964) at General Electric's Hanford Laboratories used the natural uranium in biotite to make it serve as a neutron dosimeter. Similarly, we at General Electric in Schenectady used microscope-slide glass with 0.35 parts per million of uranium to measure thermal neutron doses in reactors (Fleischer *et al.*, 1965b). The linearity of track density with time in a nuclear reactor at a fixed position for exposure to thermal neutrons (Figure 6-11) shows the usefulness as a dosimeter. Such dosimeters provide the third measurement that is normally done in fission track dating, where fossil tracks and induced tracks in the sample being dated need to be accompanied with a measure of the neutron dose used to induce tracks. One virtue of glass is that almost any amount of uranium oxide can be dissolved, so that even for low neutron exposures track densities can be high enough that counting is easy. Thus different glass dosimeters may be used where there are widely different expectations of neutron exposures.

Nuclear Weapons

Neutrons from atomic bombs might be expected to induce fission from trace amounts of uranium in glass that was near where nuclear explosions have occurred (Fleischer, 1987). Attempts to find such tracks so far have failed because appropriate samples have not been saved (or if they have been, they have not been located).

Neutron exposures at Hiroshima are still highly uncertain, and yet the estimates of exposures there are used internationally to regulate exposure to the public. Among the many forms of glass that existed at Hiroshima in 1945, the ones that are saved as souvenirs tend to be heat-affected, and consequently in most such samples the tracks are annealed out. The outwardly more ordinary remnants of glass objects tend to have been discarded. It is unfortunate that scientific realization that glasses can serve as dosimeters came 20 years after the event we would like to characterize better.

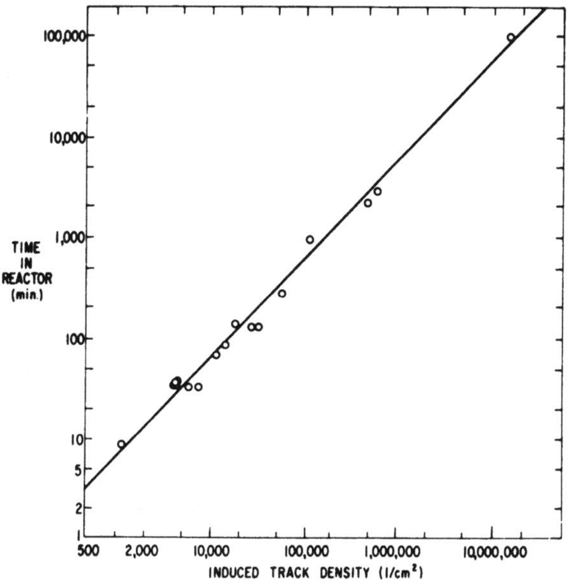

Figure 6-11. Densities of induced fission tracks, as counted in a glass dosimeter, plotted against time spent in the reactor. The linear increase in the track density with time spent in the reactor may be converted to a direct measure of the integrated neutron flux (Fleischer *et al.*, 1965b).

Professor Sakae Shimizu, who has been most helpful in my search for glass that was irradiated at Hiroshima, was one of the first scientists to enter the area after the explosion. At that time glass insulators for phone wires were held onto wooden pegs on telephone poles by fused sulphur. He used induced activity in the sulphur (stable ^{34}S being converted into radioactive ^{35}S) to measure neutron doses. Unfortunately, in a combined scientific disaster and administrative blunder, Shimizu's highly pertinent data were confiscated by an official of the U.S. occupation force, and then lost. The glass insulators, which were discarded, are superb candidates for counting neutron-induced fission tracks, but that possibility was unknown at the time.

Nuclear Safeguards

A more successful endeavor that used dosimetry based on induced fission is shown in Figure 6-12, an unattended nuclear safeguards device designed to record and characterize the use of a nuclear reactor by recording the energy spectrum of neutrons as a function of time (Weidenbaum *et al.*, 1970). A motor winds a plastic track-recording tape past small plates of ^{235}U, ^{238}U, ^{237}Np, ^{232}Th, and ^{252}Cf. Among the first four, fission induced by different energy neutrons gives a four-parameter energy spectrum. The ^{252}Cf provides a constant flux of spontaneously emitted fission fragments, and thus gives a record that the tape has moved continuously (or alternatively, has been tampered with). The device was created by General Electric's Vallecitos Laboratory for the Arms Control and Disarmament Agency.

Lightning and Neutrons

Lightning, such as is pictured in Figure 6-13, has been suggested as a dramatic source of neutrons that might explain anomalously young radiocarbon (^{14}C) ages found at some high-altitude locations (Libby and Lukens 1973). Cosmic-ray-generated neutrons in the atmosphere produce ^{14}C from ^{14}N, and an excesses of ^{14}C would result if lightning were to increase the atmospheric ^{14}C inventory locally. The devices shown in Figure 6-14 allowed a check on this proposal (Fleischer, 1975) – and made me dizzy in the process. Pictured are combined lightning counters and neutron dosimeters.

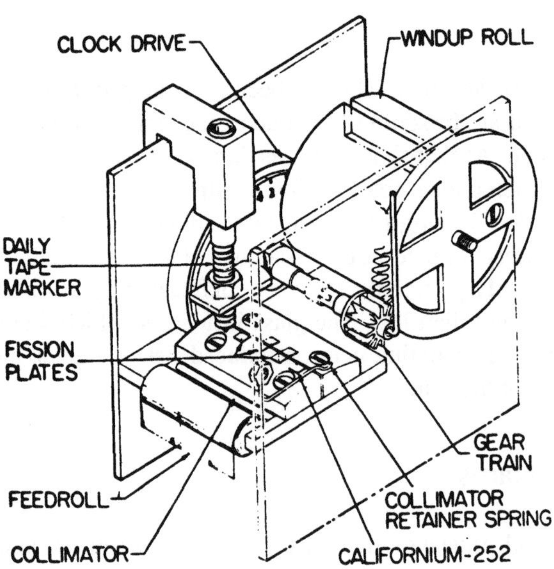

Figure 6-12. Nuclear safeguards device to monitor the use of nuclear reactors. The neutron energy spectrum is recorded as a function of time by fission induced in ^{235}U, ^{238}U, ^{237}Np, and ^{232}Th and recorded on a moving polycarbonate tape. A ^{252}Cf fission source proves a known flux of particles to permit calibration of the tape speed (Weidenbaum *et al.*, 1970).

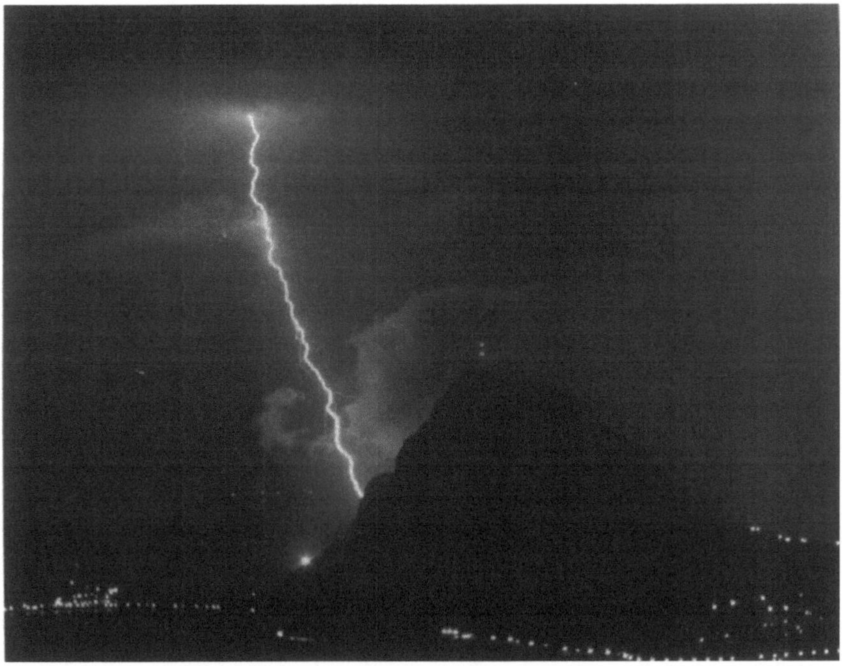

Figure 6-13. Nighttime lightning at Mt. San Salvatore, Switzerland (photo courtesy of the National Center for Atmospheric Research).

The dosimeters (like those in the safeguards device) provide information on the neutron energies, in this case using B, Pb, and Th as nuclides that react to neutrons. These dosimeters are mounted on metal caps that cover the tips of lightning rods at the top of a high structure that is frequently struck during the thunderstorm season, the location in this case being the 300-m-high antenna of television station WCIA on the plains near Champaign, Illinois. Each lightning strike leaves a clearly visible melted and resolidified region, as shown by the before-and-after appearances in Figure 6-14. A maintenance person periodically climbs the tower, and the author's dizziness was the result of observing the installation from safely on the ground, while flat on his back using binoculars. A low count of neutrons was detected, but far too few to explain the ^{14}C results. The flux corresponded to the known cosmic ray background value of 10^{-2} neutrons cm^{-2} sec^{-1} with energy above 1 MeV.

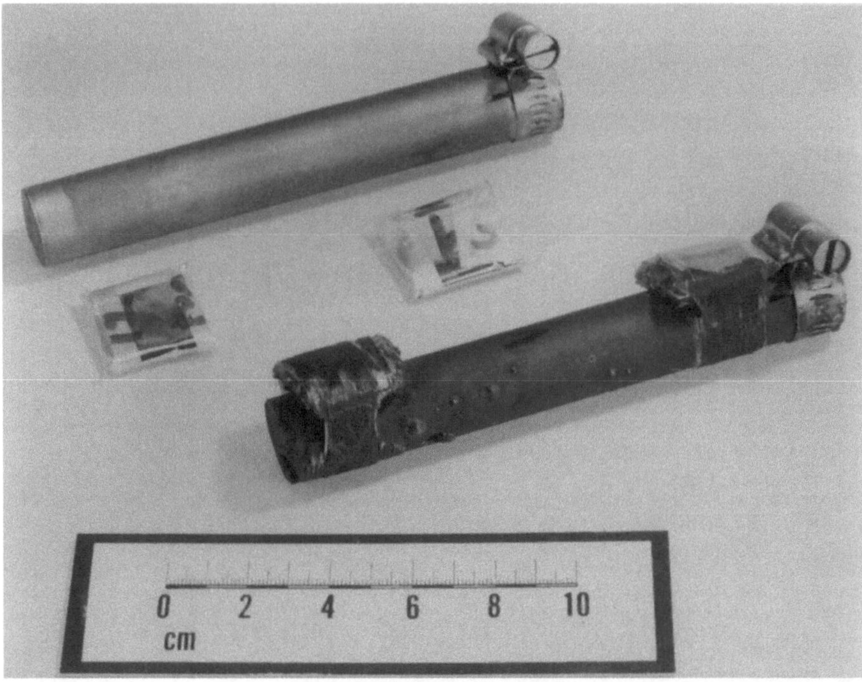

Figure 6-14. Lightning-recording caps and neutron detector packages for exposure on lightning rods 300 m above the ground. The one above shows the starting cap. The assembly below has recorded lightning strikes over a 219-day period in the form of small protrusions from the melting and resolidification of the copper.

Heavy-Ion Dosimetry

In Chapter 5 an example of direct counting of heavy cosmic ray particles was described, using Apollo helmets to detect particles to which the astronauts were exposed during missions in space. Such particles, although individually heavily ionizing, are not a hazard at ground level on earth because they are either stopped or broken up into smaller, less intensely ionizing components in the atmosphere.

For supersonic air transport, however, they are not negligible. Fukui *et al.* (1969) found that at the typical flight altitude of 20,000 m, a few percent of those heavy ions that exist in more distant space have survived their plunge into the atmosphere. They are thought to be unimportant to the occasional passenger in such aircraft, but effects would accumulate for regular crew members.

USES OF ALPHA-PARTICLE DETECTION

This section is a catch-all for some uses of tracks that are not easy to categorize. The technical uses of tracks multiplied and advanced in three main stages, first using fission tracks to learn that track etching can be done in many materials, next finding that alpha particles can be recorded in various plastic detectors, and most recently learning how to identify individual particles.

Alpha particles are important and useful since they are produced by a variety of common radionuclides and are emitted in a number of simple nuclear reactions. We already have discussed radon detection and boron mapping – in both of which alpha tracks play the major role. This chapter concludes with examples of special types of imaging, and of measurements from the bottom of the ocean and in the atmosphere.

Imaging

I now describe two examples of imaging with alpha particles that were provided by Vallecitos Atomic Laboratory. See S90 for a diversity of other imaging applications.

Figure 6-15 shows a sectioned mixed-oxide nuclear fuel rod as viewed directly (top), as imaged by conventional radiography (middle), and as seen by mapping the natural alpha particles that are emitted far more copiously by ^{239}Pu than by uranium (bottom) (Davies and Darmitzel, 1965). In conventional radiography, gamma rays from the depths of the sample reach the photographic plate and degrade the resolution. Because alphas move only short distances, the alpha radiography, done by placing the sample on a sheet of cellulose nitrate detector, records only the near-surface alpha activity; and the track detector ignores the other radiation. In the bottom view, distinct circular regions of PuO_2 in the center of the rod reveal that melting had occurred.

Neutron radiography can be done by passing a beam of thermal neutrons through an object and recording neutrons that are not scattered or captured. A piece of cellulose nitrate is placed on a layer of boron, which reacts to shoot ^4He and ^7Li particles into the detector. Figure 6-16 is such a radiograph, made of a familiar object, the light areas being the uninterrupted beam (C. Porter, see FAFPW72). Neutrons are special in often reacting with light elements such as wood or rubber, as shown here.

Note that X-rays would not be useful in showing the low-density material, as I once experienced. I ingeniously acquired a 3-cm-long wood splinter in my foot, and medical X-rays gave no hint of its presence. Its size was something of a surprise to the surgeon who ultimately removed it, and to me.

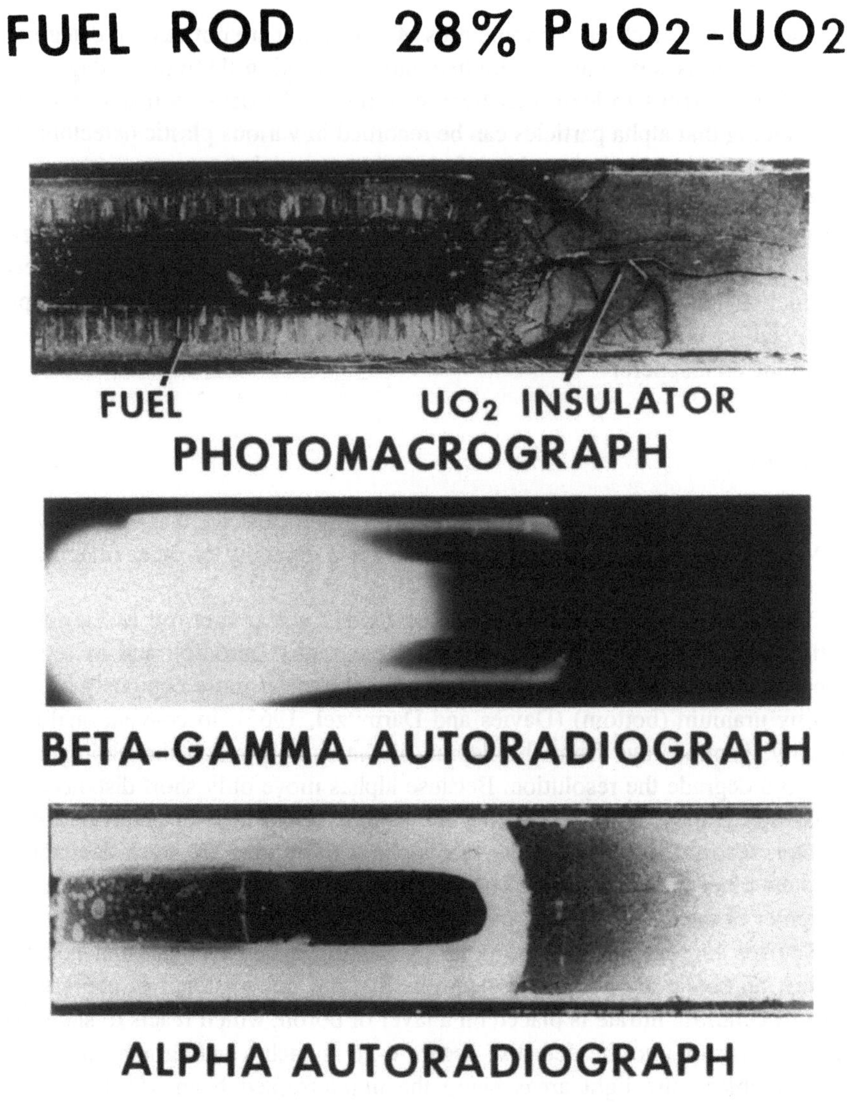

Figure 6-15. Photos of an intensely radioactive mixed oxide UO_2-PuO_2 fuel rod from an experiment in nuclear fuel development. Top: Ordinary metallographic photograph of sectioned rod. Center: beta-gamma radiograph fogged by radioactivity. Bottom: alpha-radiograph by track etching showing clearly the plutonium-rich regions as light-scattering regions that make clear where the fuel was melted (Davies and Darmitzel, 1965).

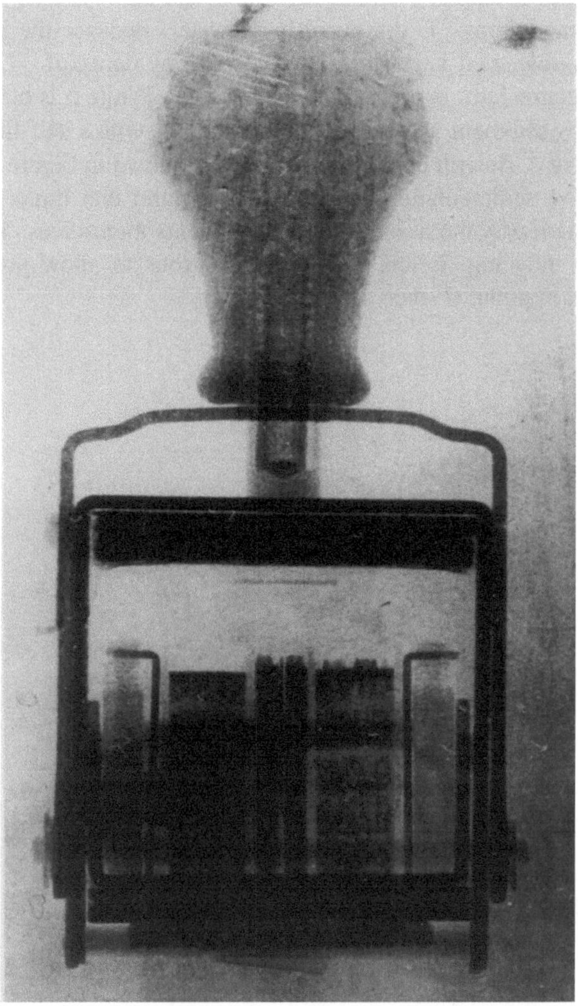

Figure 6-16. Neutron radiograph of a rubber date-stamper recorded by alpha-particle tracks in a cellulose nitrate sheet with an adjacent boron activation plate that supplied alpha particles from neutron-induced alpha-particle reactions (photo by C. R. Porter, Vallecitos Atomic Center; FAFPW72).

Sedimentation Rates in the Ocean

Radioisotopes that are formed in ocean water can be used to measure many oceanic processes. For example, sedimentation rates can be measured by means of ^{230}Th that is formed from ^{238}U dissolved in seawater. Because the solubility of thorium is negligible, as it is produced in the water by decay of ^{234}U, it precipitates and thus accumulates in the sediments in excess. While it is being progressively covered by sediment, the ^{230}Th gradually decays, with a half-life of 77,000 years. A profile of ^{230}Th with depth, such as the one shown in Figure 6-17, therefore measures the sedimentation rate. The background that limits such measurements is provided by the uranium in the sediments themselves. The uranium is measured by inducing fission tracks with neutrons to allow point-by-point determination of uranium (Fisher, 1977).

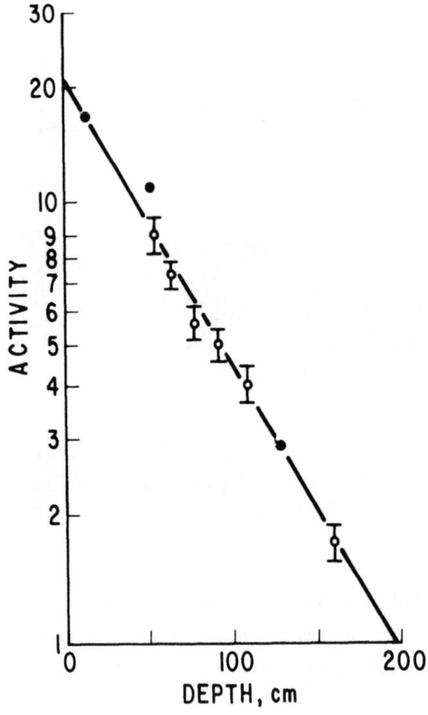

Figure 6-17. The sedimentation rate near the East Pacific Rise has been shown to be about 6 mm/1,000 years by measuring ^{230}Th, which forms from ^{238}U dissolved in the seawater. Alpha-active ^{230}Th precipitated from the seawater is recorded on plastic detectors as a function of depth in the core. The filled circles are comparison determinations made by laborious radiochemical techniques (Fisher, 1977).

Bird Altimetry

The average technical person, when told that alpha particles form tracks in plastics, is unlikely to say, "Great. Now I can measure the flight altitude of birds." Yet Kristiansson and co-workers did just that. First (Kristiansson *et al.*, 1977) they devised an integrating barometer, as sketched in Figure 6-18. The device consists of a small point-like source of alpha particles (in this case ^{210}Po) and a slanted detector of alpha tracks. The source and detector are positioned so that the alphas reach the nearest portion of the detector when the bird is at ground level. At higher altitudes, the lower atmospheric density allows the particles to make tracks farther along the detector. Thus, the distribution of the distance of tracks along the detector maps the distribution of times spent at different altitudes.

For swifts Gustafson *et al.* (1977) used a 0.4-gm device that was 2 cm in length, attached to the bird with biodegradable paste. They found that in homing flights over long distances, these birds flew at heights of up to 3,600 m in clear weather and at lower altitudes during cloudy periods. Individuals tended to resume the same altitude on flights under similar weather conditions, but different individuals chose different altitudes.

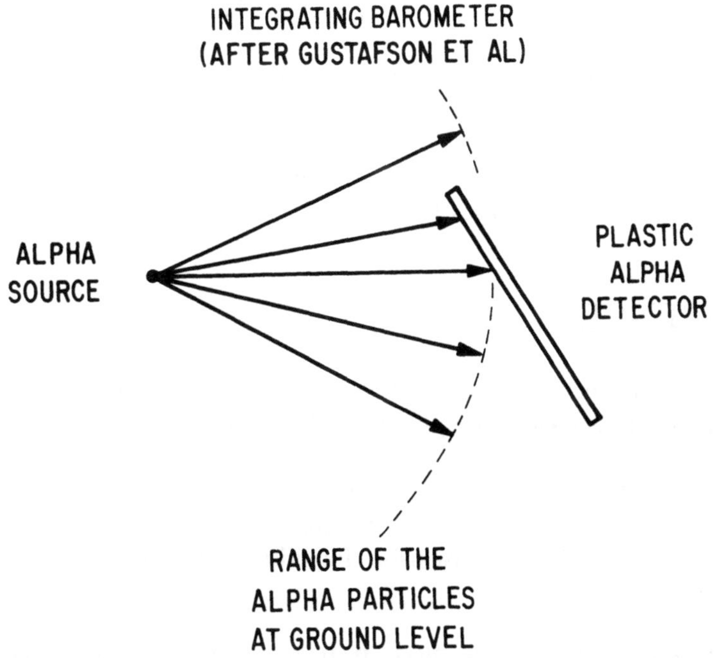

Figure 6-18. Integrating barometer. The tracks from an alpha source will be recorded at distances along a detector that increase as the pressure of the intervening air decreases.

CONCLUDING NOTE

With Chapter 6, we end the description of the science of nuclear track etching. Since the literature on this subject includes about 8,000 papers, it should be evident that I have only been able to give a sampling of the abundant uses of etched tracks.

FURTHER READING ON APPLICATIONS

Radiation Dosimetry

DB87 Chapter 7 of *Solid State Nuclear Track Detection*, S. A. Durrani and R. K. Bull, Pergamon Press, Oxford, 1987, pages 144 to 198.

Ancient tracks, nuclear technology, and others:

FPW75 Chapters 6 to 10 of *Nuclear Tracks in Solids*, R. L. Fleischer, P. B. Price, and R. M. Walker, Univ. of California Press, Berkeley, 1975, pages 307 to 595.

Diverse applications:

FAFPW72 "Particle Track Etching," R. L. Fleischer, H. W. Alter, S. C. Furman, P. B. Price, and R. M. Walker, *Science* **178**, 255–263 (1972).

MRS95 *Ion Tracks in Solids* issue of *Materials Research Society Bulletin*, Dec. 1995, **20** (no. 12), pages 17 to 41.

S90 Part II of *Ion Tracks and Microtechnology*, R. Spohr, Vieweg, Braunschweig, 1990, pages 183 to 261.

REFERENCES

J. S. Armijo and H. S. Rosenbaum (1967), "Boron Detection in Metals by Alpha-Particle Tracking," *J. Appl. Phys.* **38**, 2064–2069.

C. P. Bean, R. L. Fleischer, P. S. Swartz, and H. R. Hart, Jr. (1966), "Effect of Thermal-Neutron Irradiation on the Superconducting Properties of Nb_3Al and V_3Si Doped with Fissionable Impurities," *J. Appl. Phys.* **37**, 2218–2224.

D. S. Burnett, R. C. Gatti, F. Plasil, P. B. Price, W. J. Swiatecki, and S. G. Thompson (1964), "Fission Barrier of Thallium-201." *Phys. Rev.* **134B**, 952–963.

B. Center and F. H. Ruddy (1976), "Transuranic Elements: Detection and Characterization of Aerosols with the Nuclear Track Technique," *Anal. Chem.* **48**, 2135–2139.

D. R. Chapman and H. K. Larson (1963), "On the Lunar Origin of Tektites," *J. Geophys. Res.* **68**, 4305–4358.

R. M. Chrenko (1971), "Boron Content and Profiles in Large Laboratory Diamonds," *Nature, Phys. Sci.* **229**, 165-167.

J. H. Davies and R. W. Darmitzel (1965), "Alpha Autoradiographic Technique for Irradiated Fuel," *Nucleonics* **22**, 86–87.

D. E. Fisher (1978), "Fission Alpha Particle Track Analysis: A New Geological Technique for the Measurement of Uranium, Thorium and Isotope Disequilibria," *J. Radioanal. Chem.* **38**, 477–490.

R. L. Fleischer (1975) "Search for Neutron Generation by Lightning," *J. Geophys. Res.* **80**, 5005–5009.

R. L. Fleischer (1987), "An Opportunity and An Opportunity Lost," *Health Phys.* **52**, 219–221.

R. L. Fleischer and A. C. Delany (1976), "Determination of Suspended and Dissolved Uranium in Water," *Anal. Chem.* **48**, 642–645.

R. L. Fleischer and D. B. Lovett (1968), "Uranium and Boron Content of Water by Particle Track Etching," *Geochim. Cosmochim. Acta* **32**, 1126–1128.

R. L. Fleischer and P. B. Price (1964), "Fission Track Evidence for the Simultaneous Origin of Tekitites and Other Natural Glasses," *Geochim. Cosmochim. Acta* **28**, 755–760.

R. L. Fleischer and O. G. Raabe (1977), "Fragmentation of Respirable PuO_2 Particles in Water by Alpha Decay – A Mode of 'Dissolution'," *Health Phys.* **32**, 253–257.

R. L. Fleischer, C. W. Naeser, P. B. Price, R. M. Walker, and M. Maurette (1965a), "Cosmic Ray Exposure Ages of Tektites by the Fission Track Technique," *J. Geophys. Res.* **70**, 1491–1496.

R. L. Fleischer, P. B. Price, and R. M. Walker (1965b), "Tracks of Charged Particles in Solids," *Science* **149**, 383–393.

R. L. Fleischer, P. B. Price, R. M. Walker, and E. L. Hubbard (1966), "Ternary Fission of Heavy Compound Nuclei in Thorite Track Detectors," *Phys. Rev.* **143**, 943–946.

R. L. Fleischer, H. R. Hart, Jr., and G. M. Comstock (1971), "Very Heavy Solar Cosmic Rays: Energy Spectrum and Implications for Lunar Erosion," *Science* **171**, 1240–1242.

R. L. Fleischer, L. G. Turner, H. G. Paretzke, and H. Schraube (1984), "Neutron Dosimetry Using Particle Tracks in Solids: An Intercomparison," *Health Phys.* **47**, 525–531.

R. L. Fleischer, H. R. Hart, Jr., K. W. Lay, and F. E. Luborsky (1989), "Increased Flux Pinning upon Thermal-Neutron Irradiation of Uranium-Doped Y-Ba-Cu-O," *Phys. Rev.* **B40**, 2163–2169.

K. Fukui, Y. K. Lim, and P. S. Young (1969), "Cosmic-Ray Heavy-Nucleus Enders at Various Atmospheric Depths," *Nuovo Cim.* **61B**, 210–219.

T. Gustafson, B. Lindkvist, L. Gotborn, and R. Gyllin (1977), "Altitudes and Flight Times for Swifts *Apus apus. L.*," *Ornis Scand.* **8**, 87–95.

W. S. S. Jee, R. B. Dell, and L. G. Miller (1972),"High Resolution Neutron-Induced Autoradiography of Bone Containing ^{239}Pu," *Health Phys.* **22**, 761–763.

J. D. Kleeman and J. F. Lovering (1967), "Uranium Distribution Studies by Fission Track Registration in Lexan Plastic Prints," *At. Energy in Australia* **10**, 3–8.

K. Kristiansson, B. Lindkvist, and T. Gustafson (1977), "An Altimeter for Birds and Its Use," *Ornis Scand.* **8**, 79–86.

L. M. Libby and H. R. Lukens (1973), "Production of Radiocarbon in Tree Rings by Lightning Bolts," *J. Geophys. Res.* **78**, 5902–5903.

A. R. Moorthy, C. J. Schopfer, and S. Banerjee (1988), "Plutonium from Atmospheric Weapons Testing: Fission Track Analysis of Urine Samples," *Anal. Chem.* **60**, 857A–860A.

P. B. Price and R. M. Walker (1963), "A Simple Method of Measuring Low Uranium Concentrations in Natural Crystals," *Appl. Phys. Lett.* **2**, 23–25.

P. Prevo, R. E. Dahl, and H. H. Yoshikawa (1964), "Thermal and Fast Neutron Detection by Fission-Track Production in Mica," *J. Appl. Phys.* **35**, 2636–2638.

R. Scanlon (1975), unpublished. See p. 574 in FPW75.

R. M. Walker, P. B. Price, and R. L. Fleischer (1963), "A Versatile Disposable Dosimeter for Slow and Fast Neutrons," *Appl. Phys. Lett.* **3**, 28–29.

R. Weidenbaum, D. B. Lovett, and H. D. Kosanke (1970), "Flux Monitor Utilizing Track-Etch Film for Unattended Safeguards Application," *Am. Nucl. Soc. Trans.* **13**, 524–526.

Chapter 7
LINKS OF SCIENCE AND TECHNOLOGY

For every complex problem there is a solution
that is simple, neat, and wrong.
H. L. Mencken

Science is many things to different people – a quest for deeper knowledge of the world around us, a source of vast enjoyment, and a means of serving mankind through innovation. In the earlier chapters I have tried to make clear each of these roles of nuclear tracks. Here we examine the interplay of science and technology, starting with the deliberate purpose of translating newly recognized science into practical, commercial accomplishments.

To this purpose the Research Laboratory held the first conference on etched nuclear tracks – a purely General Electric affair. It was a single-day event attended by 44 scientists and engineers on 15 June 1965 at the Research Laboratory in Schenectady, New York – The Particle Track Etching Information Meeting. The track studies begun in Schenectady only four years earlier had shown promise in a wide variety of technologies, and applications were being pursued or considered by an astonishing 13 components of the company. Three talks summarized the basic science as it was then known; nine talks covered applications that ranged from filters and uses of micropores to imaging techniques, tracers, uranium prospecting, dosimetry, and elemental distributions in nuclear fuels and fuel cladding (see Fleischer, 1997).

This burst of divergent science and technology was the result, I believe, of the conjunction of two factors. The first was finding stable, chemically etched tracks in a broad range of materials, first in micas by Price and Walker (1962a) and soon thereafter in plastics, glass, and diverse minerals by Fleischer and Price (1963a,b; 1964a). The other factor was that the discoveries occurred in a company that then had an abundance of technological interests – enhanced by a desire to build new businesses. A second corporate influence was the encouragement of basic science at the Research Laboratory – with the expectation that some of it would turn out to be of economic benefit to the company (see the quotation on page v by Suits and Bueche, SB67). For tracks the expectation was met, but this policy of the 1960s has since vanished. Nevertheless, these discussions will emphasize the interconnections of science and technology that began in the 1960s.

Incidentally, the title of this chapter is the same as that of a series of articles initiated by the innovative editor of the MRS Bulletin of the Materials Research Society – E. L. Fleischer – to describe connections of science to practical advances in materials . See for example, the first such article, by Briant and Bewlay (1995). It rarely is a severe detriment to let those responsible for paying for science know that its results can be of practical use. I now summarize some of the practical results from track etching and then ask some questions about tracks and their relevance to technology and science.

APPLICATIONS

Most of the areas of practical application that were contemplated in the 1965 track conference continue to be of considerable interest and activity, but noticeably absent at the meeting was what probably is of the greatest current economic importance – measuring the partial fading of historical fission tracks. Track dating was known then (Price and Walker, 1963) and had been demonstrated (Fleischer and Price, 1964b; Fleischer *et al.*, 1964a; Maurette *et al.*, 1964), but the potential practical uses of thermal effects on tracks were not yet apparent in 1965. Establishing a fission-track-dating operation was considered, but rejected as having an uncertain future and being distant from General Electric interests.

Business Derived From Science

The commercial applications that have led to companies that depend on track etching are listed in Table 7-1 (Fleischer, 1997). Other business both contribute to the etched track work and profit from it, one group by supplying detectors and another group by selling image-analyzing microscopes to facilitate track counting and related track measurements.

The first recognized practical result from track etching was hole engineering. As noted in Chapter 2, track etching of thin detector sheets produces holes of a single size that are specified by the duration of etching; and the number of holes is controlled by the dose of particles that make tracks. Filters with pores having diameters from 10^{-6} to 10^{-3} cm can be produced. Such sieves have found use in cytology (since holes can be the size of biological cells), aerosol monitoring, oceanographic particulate sampling, and in producing ultraclear solutions for the semiconductor industry. Draft beer that can be stored without refrigeration can be made by sieving out the bacteria and yeast. The filters were produced by General Electric's Vallecitos Atomic Laboratory as Nuclepore Filters, an operation that later became Nuclepore Corporation, which now is part of Corning

Scientific Products Division. Other companies have entered the market, which by one (perhaps generous) estimate (Kuznetsov *et al.*, 1995) is around $30 million a year.

Table 7-1. Origins of Commercial Uses of Etched Particle Tracks

Scientific observation	Commercial application	Notes
Track etching makes holes	Controlled holes; filters	Cleaning solutions, sampling air and water for particles
Alpha particles make tracks	Radon measurements in homes and nature	Radiobiological dosimetry, mineral resource exploration, geological hazard prediction
^{238}U fissions, and tracks are stored	Ancient heating of rocks can be measured	Avoid drilling for oil where heating was excessive
Neutrons cause recoil tracks and induce reactions	Neutron dosimetry	Radiation dose measurements

The second principal commercial use of track detectors is measuring radon in homes and in nature, as was described in Chapter 3. Measurements are possible because alpha-particle tracks can be recorded in plastic detectors (Fleischer *et al.*, 1964b). Again multiple applications were pioneered at Vallecitos, which later transferred the endeavors into a separate firm, Terradex Corporation, now part of R. S. Landauer. The original use in uranium exploration has diminished in the United States, but measurements in homes simultaneously expanded, driven by public alarm about indoor radioactivity. Such worries began around 1980 in Sweden and were strongly driven in the United States in 1985 by publicity about the level of radon in the Watrous house, as noted at the opening of

Chapter 3. Terradex earned about $2 million in 1985, and was sold in 1986 at $47 per share, (the original cost to the founders of the firm was $1 per share in 1974). In the aftermath of the Watrous publicity, Landauer found that they needed to more than double the number of employees in the radon work, and they were selling more than $3 million per week of radon tests. Sales have now steadied at a much lower level – a few million dollars per year. Many other firms have entered the business of track-based radon monitoring, making it difficult to assess the total current annual income.

The most recent of the developments listed in Table 7-1 is the use of thermal effects on spontaneous fission tracks in minerals to learn about ancient heating. Deciphering these messages in the "tracks of time" often allows petroleum geologists to decide whether to abandon a site or to probe it further. Given the immense cost of deep drilling, which can be many millions of dollars per hole, the potential savings are of major importance. This use of tracks is probably the single largest commercial benefit of track etching. Note that the savings to the oil companies far exceed the costs of the track measurements that are used to make informed decisions on drilling. One first-order estimate is that oil companies over the last decade have spent about $10 million on thermal age measurements at a price of typically $1,000 each. In contrast, if the drilling of 500 deep holes has been avoided, there is a saving of perhaps $30 billion, a noticeable return on investment.

Neutron dosimetry, i.e. measuring personal neutron doses, (R.S. Landauer, Neutrak detectors) is another track business that has existed since about 1978. At 200,000 measurements per year, their gross income would be roughly $1.3 million.

Other Technological Applications

Numerous "practical" uses of tracks that have not acquired the status of becoming business operations, but nevertheless serve the good of mankind, are listed in Table 7-2, which is adapted from Fleischer (1997). These are technological applications that aid commercial or governmental processes or products, but the magnitudes of investment are awkward to calculate. Even though the dollar values of these applications are difficult to decide, such improvements may make competitive differences that allow products to continue and flourish in the marketplace.

The dosimetry and radiobiology entries for example are supportive to intelligent radiation protection, and the materials-development and superconductors items aid in industrial processing to optimize products, producing better color in diamonds, improved currents in superconductors, and better magnetic and corrosion properties in steel. More details of many of these applications are described in FPW75, S90, and MRS95.

Table 7-2. Other Technological Applications of Etched Particle Tracks

Area	Application
Atmospheric sciences	Detection of nuclear weapons tests by fissionable radionuclides in aerosols
	Measuring the flight altitudes of birds
Dosimetry	Counting heavy ions in space missions and supersonic aircraft
Fusion	Mapping alpha emission in fusion reactors
Imaging	Neutron radiography of low-density objects using boron or uranium fission plates
	Alpha radiography of nuclear fuel
	Heavy-particle radiography
Materials development	Controlling color of diamonds using boron or nitrogen mapping
	Improving boron distribution and resultant properties in magnetic steels and stainless steels
	Locating fissionable additions in superconductors
	Improving information storage in magneto-optic films
Nuclear technology	Fission rate measurements aid reactor engineering
	Measuring neutron energy spectrum at reactors
Particle counters	Conductivity through a narrow single track-etched hole characterizes particles in electrolytes
Radiobiology	Mapping fissionable atoms in tissue
Superconducting materials	Improving high-field current capacity by radiation damage from fissioning deliberately added uranium
Tracers	Alpha and fission properties characterize designed particles

SCIENCE ALSO BENEFITS FROM TECHNOLOGY

Just as technology often benefits from science, it also contributes to science. Table 7-3, adapted from Fleischer (1997), lists some examples.

Table 7-3. Technological Aids to Science

Technology	Enabled science
Synthetic mica	Stabilizing tracks for transmission microscopy
Hole detection method for capacitor materials	Rare, superheavy cosmic rays could be located
Filters	Test theory of one-dimensioned superconductivity
	Measure size distribution of cloud nuclei
	Red cell motions in restricted passages
	Test ionic mobility and diffusion on a molecular scale
	Test electric-field effect on superfluidity of liquid helium
	Create quantized vortex rings in superfluid helium
	Test model of phonon scattering
Tests of radiation effects on TV camera on the moon	Measure lunar surface erosion rates
Use of polycarbonate for space helmets	Measure heavy cosmic rays encountered at the moon
Controlled vacuum leak	Etching tracks to optically visible size (to adjust flow)

Track-based filters were the technological result of scientific exploration. But they in turn have enabled many scientific studies. The subjects range from superconductivity in minute wires formed in pores (Possin, 1971) to measuring cloud-forming nuclei separated from the air by filters (Twomey, 1972) deformability of red cells (Gregersen *et al.*, 1967), diffusion in solutions (Varro *et al.*, 1978) [sometimes with pores of molecular dimensions (Beck and Schultz, 1972; Quinn *et al.*, 1972)], observing electric-field effects on superfluid helium (Notarys, 1968), creating quantized vortices in superfluid helium (Gamota, 1973), and testing phonon transport (Zaitlin and Anderson, 1975).

The item concerning the TV camera on the moon would be cryptic without further explanation. The camera that was returned contained a glass filter that faced space through a protected enclosure. This glass allowed us to measure the local low-energy, heavy-ion flux at each depth in the glass, unaltered by erosional effects. The variations with depth in lunar samples that are subjected to erosional processes are altered, but the differences between those behaviors and that in the glass allows the rates of erosion to be found down to ~0.1 nm per year.

The first item is particularly momentous in Table 7-3, as we noted many years ago (FPW75). Synthetic mica had been made earlier and the guy down the hall at G.E. (Louis Navias) had some. Tracks were stable in that mica – in contrast to the fading observed in natural mica in the electron microscope, a behavior that ended Silk and Barnes' (1959) activity with tracks. Navias' mica was a key in providing lasting tracks. It thus enabled Price and Walker's early experiments on tracks to progress (Price and Walker, 1962b). And this step nucleated the track program that expanded so strikingly at General Electric.

ARE INTERCONNECTIONS HELPFUL?

Interconnections are critical. As both science and technology become more diverse, two possibilities arise, each of which will exist in some measure. One is that more isolated pockets of workers in specialized areas will exist, because there are more specialties. The second possibility is that more points of contact will occur between workers in different subdisciplines. The first is an obvious hazard: Discoveries don't propagate outside of their original context because the folks in one lab don't know what the guys next door are doing, or don't care. The second presents many opportunities, and Tables 7-1, 7-2, and 7-3 constitute a compendium of examples for the field of track etching. A wealth of new opportunities can arise when people with different knowledge and interests come together. And the diversity both within the General Electric Research Laboratory and through disparate parts of General Electric ensured that the boundaries of separate areas of knowledge came together promptly after track etching was begun. The positive results are included in the tables. It must be emphasized that a vast outpouring of science and technology has evolved that is outside General

Electric – an estimated 250–300 publications per year since about 1970 (Fleischer, 1991), and many commercial endeavors, for example those in dosimetry, thermal dating, and uranium exploration. But the situation within General Electric was special in the way that it facilitated the prompt exploration of the many scientific and technological uses of etched tracks.

The directions of the interconnections are diverse and, as Figure 7-1 indicates, sometimes unpredictable. The tables show that some transitions progress from science to technology and some reverse the direction. In the case of filters, Table 7-3 contains multiple examples where science originally led to the technology of filter production, which in turn enabled multiple basic scientific works. Once or twice I have tried to sketch a simple diagram to make graphic the flow of innovation and progress in nuclear track etching. The result looked much like a complex, chaotic spider web, such as is created by a spider that is under the influence of a drug-stimulated high (Noever *et al.*, 1995), as shown in Figure 7-2. Some interconnections were natural and obvious; others were startlingly unexpected. Two examples of the latter are 1) the connection between the effects of heat on nuclear tracks and knowing when to drill for petroleum (Chapter 4) and 2) that between the registration of alpha particles and the flight altitudes of birds (Chapter 6).

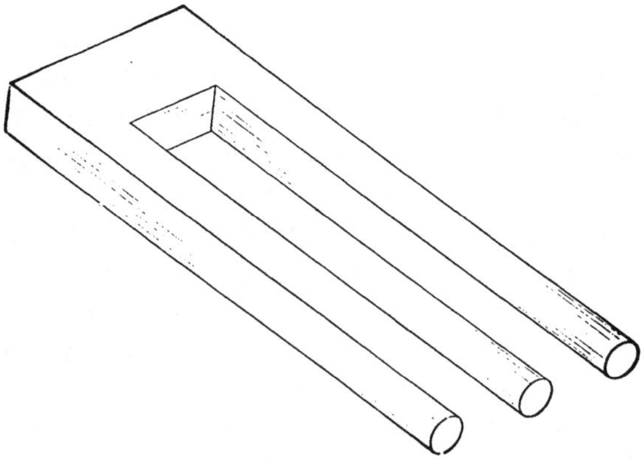

Figure 7-1. Interconnections – a sketch with a lesson: From the way things begin, it is often not obvious how they will end.

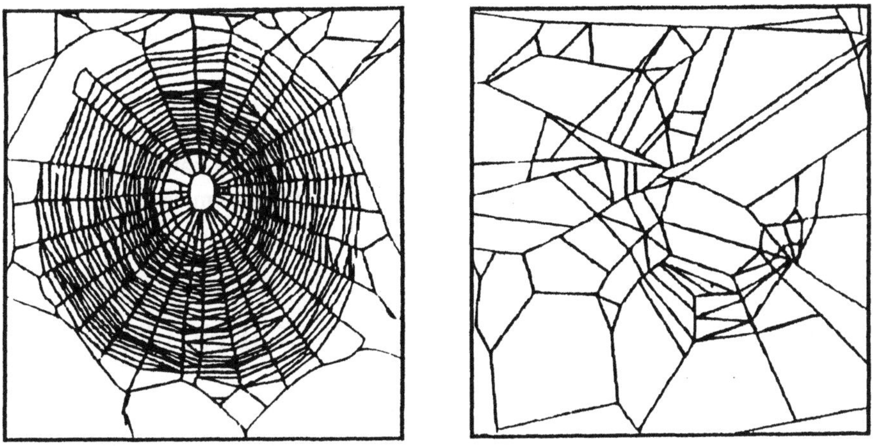

Figure 7-2. The orderly connections of a conventional spider web (left) are jumbled (right) if the spider is drugged – by caffeine in this case (Noever *et al.*, 1995). Connections of science and technology have similar disorganized intricaces.

INDUSTRIAL SCIENCE: WHERE WOULD TRACKS HAVE BEEN WITHOUT IT?

The question in this heading goes beyond science. It demands speculation for a response. Would practical uses of tracks have expanded if a diverse company were not involved? I expect that many present-day ideas would probably ultimately have developed, but far more slowly. The closest approach to evidence that I can cite is that neither of the early separate pioneering works of Young (1958) and Silk and Barnes (1959), which were not done in industry, stimulated further inquiry into the nature of tracks or their scientific and technological uses. Four years later, work began at General Electric, and suddenly, three years after that, the conference that was described at the start of the chapter took place. This event showed that most of the main directions that continue to occupy track people had been recognized as the result of a determined focus on practical opportunities – work by many interested people within more than a dozen parts of G.E.

Note, however, that the testing of many practical paths for track etching was not a spontaneous outpouring from the various businesses within General Electric. Rather, a specific operating unit, Vallecitos Atomic Laboratory of the Nuclear Energy Division, was given responsibility for exploring possibilities. People there recognized new applications, organized assessment of most of the practical areas, and designated who should look into the various opportunities.

The rapid beginnings of commercialization at General Electric allowed the company to attain a useful patent position – consisting of more than two dozen that have been issued. Patents can be particularly important at the start of a business, encouraging investment and providing some sense of security during an otherwise difficult, albeit exciting, time. Because patents have a designated, limited life, slow development of a business can dissipate much of the advantage that they offer their owners.

HOW HEALTHY IS INDUSTRIAL SCIENCE?

The term *industrial science* as applied to diverse industrial firms in the United States has become a virtual oxymoron. I was in a direct position to observe how General Electric led the way in turning off its role in science, starting in 1965, but progressing at an accelerating rate in the years following 1980. During the early 1990s decreases in research and development also became evident at AT&T and IBM (May, 1996). Even in Japan research in the private sector is said to be decreasing (Hayashida, 1996).

Why Should There Be Science in Industry?

The premier justification for undirected science in industry is the serendipitous discovery of new practical possibilities that otherwise would not be identified, or whose recognition would come much later. This philosophy for "speculative ventures in research" was expressed by former General Electric Directors of Research Suits and Bueche (1967): "Although existing needs and problems dictate the largest part of research in industry – or in the Nation, for that matter – experience has shown repeatedly that studies begun solely to grapple with fundamental questions of science can be rewarding to industrial technology, providing advances for which there was no recognized pre-existent need, however strong that 'need' might become afterwards." Note that "undirected science" is not really without managerial guidance, since it is understood to be in areas of potential use to the company

George Wise has added in a personal communication that one major type of success from doing science in industry has been to equip companies to be prompt, able followers – using initial advances made by others – nuclear magnetic resonance imaging being a recent example at General Electric.

Other familiar reasons why industrial science was beneficial to businesses were noted by Fowler (1995). "Such work enhanced the reputation of the company; it made university relations and recruiting of top-quality scientists easier; and it provided in-house consulting of the highest caliber."

Fowler's three items are subsidiary reasons why industry benefits from their own science. National science and technology gain in another way – through preparation of future professors, who will aid in training other generations of scientists and engineers, and will do so after having gained some significant knowledge of the practical needs of industry and of society.

As an example, from a science group that consisted of 26 technical staff members of which I was one at General Electric in 1965 – the Physical Sciences Branch – more than half, 14, left for universities. Of the 236 Ph.D. members of the staff that were doing technical work in November 1961, I could find the subsequent activities for 228. Of those, 83 (36%) ultimately entered universities. Half of these were teaching prior to age 44 and 70% by age 55. This cadre constitutes a major imput of technological realism in training young people for future careers in science and engineering. And G.E.'s research establishment was by no means the largest in the United States at that time. Hence, the total contribution of former industrial scientists to teaching in science and engineering departments in universities during past decades must be considerable. Assessing its overall magnitude quantitatively would be an interesting undertaking.

The people who entered university life were also active contributors to recognized science, as evidenced by the citations they received. Of the Research Laboratory group just described, a subgroup of 36 who migrated to 12 major, research-oriented universities received an average of 53 literature citations in the fifth year after they took up their university positions. For comparison, in 1968 in physics at the University of Illinois – a highly regarded department – the full professors averaged 14 citations.

Did General Electric Gain From Etched Tracks?

Aside from the considerations quoted from Fowler (1995), what did G.E. gain from its work on track etching? We have listed two businesses, Nuclepore filters (Chapter 2) and Terradex radon monitoring (Chapter 3); contract-supported work having to do with nuclear weapons (Chapter 6); and process and product improvements listed in Table 7-2.

Although there are imponderables in translating knowledge of these activities into quantitative cash flow, from the estimates given earlier it is improbable that the company grossed as much as $100 million per year. For perspective, a single business of that size is regarded within G.E. as a small one, and General Electric has a well-earned reputation for not operating small businesses successfully. It is difficult to convince a high-level manager whose future depends on the success of a $500 million per year business that he should devote serious time to a far smaller operation. In addition, a large company imposes cumbersome formalities that obstruct the flexibility and ability for quick responses that are advantages of a small business. These considerations were factors in the decisions to separate Nuclepore and Terredex into partially owned, independent firms and ultimately to sell General Electric's interests in them.

The company did not gain from the businesses in neutron dosimetry and fission-track dating of thermal histories – those are areas that it chose not to enter. And the greatest commercial good from tracks, it appears, is being harvested by improved drilling success of the petroleum companies in areas that G.E. did not enter (Chapter 4).

What Has Society Gained From Etched Tracks?

The balance sheet of benefits looks more favorable here: Track etching has resulted in radiation protection from radon and neutrons, filters for diverse uses, improvements in exploration for oil and uranium, improved products and processes, and the other items in the three lists of Tables 7-1, 7-2, and 7-3. And employment for workers in the various new businesses is another significant positive factor.

A DILEMMA

To the extent that the path of nuclear tracks might be characteristic of those of other areas of science in industry, they present society with a dilemma. The assertion here is that chance practical opportunities from scientific discoveries are efficiently explored in industry, but they have larger benefits to society as a whole. Yet, often, the commercial returns are only partially to the company that developed the new technology, and the balance sheet may fail to satisfy the cost-benefit ratios that business demands.

How is the desired science to be done and, where possible, used practically?

Approaches

Recently, a group of high-level executives from more than a dozen major broad-based companies in the United States signed communications urging the government to finance more university research – as a support for ideas and invention, and to train future technical employees of their companies (Allen *et al.*, 1995). To someone who observes present-day neglect of science by industrial firms, these pleadings appear, if not actually hypocritical, at least futile. Is it reasonable to expect decision-makers in government to be eager to invest in science from which industry has withdrawn? (See the quotation by Mencken on the first page of this chapter.)

T. Kealy (K96), in a recent book, has added a new dimension to the question of how countries should support science. He presents data to show that, far from expanding the total science effort, government support causes industry to "let the government do it" by diminishing their effort by *more* than the added government funding. And Kealy asserts that far more useful science is done under industrial decision-making than by governmental fiat.

If science is largely not being done in industry, and industrial science would be beneficial to society as a whole, will that work be done? Probably not. One approach is to reason as follows: Since only some of the technology that an industrial laboratory develops from its scientific work aids that particular company, but all of it can contribute somewhere in society, government should share in the support of science in industry.

In the United States the National Science Foundation provides the primary support for the physical sciences, but with a strong bias in favor of university science. The suggestion here is different. Grants to industries for scientific work in agreed-upon broad areas could be made based on matching long-term industrial commitments to do science. Since both industry and society expect benefits from a portion of such efforts, it is appropriate that both contribute. If such grants were instituted in a substantial manner, they would represent a profound change in science policy. My view is that the losses in industrial science are profound too, and thus they merit serious attention.

The same objective might be approached by granting favorable tax status to basic studies done in industrial laboratories. Such a partial return to the previous tax credits for long-term research and development may be a more palatable, but less effective, form of government support.

CONCLUDING REMARKS

In this book I tried to convey one meaningful lesson – that progress often occurs by the intersection of fields, be they scientific or engineering – workers in one area bringing fresh views to other areas. Science aids practical technology; technology brings new tools to science. The opportunity with tracks began at a crossing between solid-state physics and nuclear science first by the recognition that an immense variety of solids can serve as nuclear track chambers and second by the documentation of their properties.

The borders between engineering and science are often blurred, but occasionally clear. From scientific understanding of what particles leave tracks in minerals, fission-track dating was shown to be possible, and rock ages were measured as contributions to the earth sciences. The transition from science to engineering occurred when petroleum geologists used track thermal responses to decide where oil might exist and where it would have been destroyed. The several radon stories are applied science, but they were enabled by the progress in scientific exploration of polymeric track detectors. These are cases in which industrial investments in science fueled practical advances.

The unique opportunities of science in industry that existed in the 1960s and 1970s enabled the rapid ballooning of the field of nuclear track etching. A conclusion is that the fabric of science and technology has intermingled threads. Occasionally a clearly scientific strand led to a particular technological advance, and sometimes the case was reversed. Typically the most direct highway to practical uses of tracks crossed boundaries within and between science and technology more than once.

FURTHER READING ON LINKS OF SCIENCE AND TECHNOLOGY

A book with the iconoclastic view that government support of science decreases science industry in more than the government adds:

K96 *The Economic Laws of Scientific Research*, T. Kealy, St. Martin's Press, New York, 1996, 382 pages.

Former General Electric view on science and technology:

SB67 C. G. Suits and A. M. Bueche (1967), "Cases of Research in a Diversified Company," in *Applied Science and Technological Progress* (report to the Committee on Science and Astronautics, U.S. House of Representatives by the National Academy of Sciences), U.S. Gov. Printing Office, Washington, D.C. June, pages 297–346.

REFERENCES

W. W. Allen, N. R. Augustine, J. L. Clendenin, R. J. Eaton, G. M. C. Fisher, R. W. Galvin, L. V. Gerstner, Jr., J. T. Gorman, G. Greenwald, G. H. Heilmeier, J. R. Junkins, J. McDonnell, R. L. Tobias,P. R. Vagelos, J. F. Welch, and E. S. Woolard, Jr. (1995), "A Moment of Truth for America," *Washington Post*, May 2.

R. E. Beck and J. S. Schultz (1972), "Hindrance of Solute Diffusion within Membranes as Measured with Microporous Membranes of Known Pore Geometry," *Biochim. Biophys. Acta* **255**, 273–303.

C. L. Briant and B. P. Bewlay (1995), The Coolidge Process for Making Ductile Tungsten: The Foundation for Incadescent Lighting," *Mat. Res. Soc. Bull.* **20** (No. 8) 67–73.

R. L. Fleischer (1991), "Where Are Nuclear Tracks Leading? – Directions in Track Studies," *Nuclear Tracks and Radiation Measurements* **19**, 847–854.

R. L. Fleischer (1997), "Tracks to Innovation – Interplay Between Science and Technology," *Radiation Measurements* **28**, in press.

R. L. Fleischer and P. B. Price (1963a), "Tracks of Charged Particles in High Polymers," *Science* **140**, 1221–1222.

R. L. Fleischer and P. B. Price (1963b), "Charged Particle Tracks in Glass," *J. Appl. Phys.* **34**, 2903–2904.

R. L. Fleischer and P. B. Price (1964a), "Techniques for Geological Dating of Minerals by Chemical Etching of Fission Fragment Tracks," *J. Geochim. Cosmochim. Acta* **28**, 1705–1714.

R. L. Fleischer and P. B. Price (1964b), "Glass Dating by Fission Fragment Tracks," *J. Geophys. Res.* **69**, 331–339.

R. L. Fleischer, P. B. Price, E. M. Symes, and D. S. Miller (1964a), "Fission Track Ages and Track-Annealing Behavior of Some Micas," *Science* **143**, 349–351.

R. L. Fleischer, P. B. Price, R. M. Walker, and E. L. Hubbard (1964b), "Track Registration in Various Solid State Nuclear Track Detectors," *Phys. Rev.* **133A**, 1443–1449.

A. B. Fowler (1995), "What Has Happened to Research at Industrial Laboratories?" *Phys. Soc. News* **4** (no. 11).

G. Gamota (1973), "Creation of Quantized Vortex Rings in Superfluid Helium," *Phys. Rev. Lett.* **31**, 517–520.

M. L. Gregersen, C. A. Bryant, W. E. Hammerle, S. Usami, and S. Chen (1967), "Flow Characteristics of Human Erythrocytes Through Polycarbonate Sieves," *Science* **157**, 825–827.

H. Hayashida (1996), "Science Policy in Japan," *Science* **272**, 1567.

V. I. Kuznetsov, L. V. Kuznetsov, and V. D. Shestakov (1995), "Track Membranes of the Third Generation. (TMG-3): Their Properties and Industrial Application," *Nucl. Instr. Methods.* **B105**, 250–253.

M. Maurette, P. Pellas, and R. M. Walker (1964), "Etude des Traces de Fission Fossiles Dans le Mica," *Bull. Soc. Franc. Min. Crist.* **87**, 6–17.

M. May (1996), "The Ups and Downs of Industrial R&D," *The Industrial Physicist* **2** (No. 2, June), 16.

D. A. Noever, R. J. Cronise, and R. A. Relwani (1995), "Spiders and Crystallographers Team Up," *Adv. Materials and Processes* **149** (8), 4.

H. A. Notarys (1968), "Electric Field Suppression of the Lambda Point in Liquid Helium," *Phys. Rev. Lett.* **20**, 258–259.

G. E. Possin (1971), "Superconductivity in Nearly One-Dimensional Tin Wires," *Proc. Intl. Conf. on the Science of Superconductivity*, F. Chilton (ed.) *Physica* **55**, 339–343.

P. B. Price and R. M. Walker (1962a), " Chemical Etching of Charged-Particle Tracks in Solids," *J. Appl. Phys.* **33**, 3407–3412.

P. B. Price and R. M. Walker (1962b), "Electron Microscope Observation of Etched Tracks from Spallation Recoils in Mica," *Phys. Rev. Lett.* **8**, 217–219.

P. B. Price and R. M. Walker (1963), "Fossil Tracks of Charged Particles and the Age of Minerals," *J. Geophys. Res.* **68**, 4847–4862.

J. A. Quinn, J. L. Anderson, W. S. Ho, and W. J. Petzny (1972), "Model Pores of Molecular Dimensions: the Preparation and Characterization of Track-Etched Membranes," *Biophys. J.* **12**, 990–1007.

E. C. H. Silk and R. S. Barnes (1959), "Examination of Fission Fragment Tracks with an Electron Microscope," *Phil. Mag.* **4**, 970–971.

S. Twomey (1972), "Measurements of the Size of Natural Cloud Nuclei by Means of Nuclepore Filters," *J. Atmo. Sci.* **29**, 318–321.

T. Varro, G. Somogyi, Zs. Varga, and L. Madi (1978), "Study of Transport of Uranyl Ions in an Ion-Exchange Membrane by Solid-State Track Detector," *Intl. J. Appl. Rad. Isotopes* **29**, 381–385.

D. A. Young (1958), "Etching of Radiation Damage in Lithium Fluoride," *Nature* **182**, 375–377.

M. P. Zaitlin and A. C. Anderson (1975), "Phonon Thermal Transport in Noncrystalline Materials," *Phys. Rev.* **B12**, 4475–4486.

FURTHER IN-DEPTH READING

Books with broad coverage of etched tracks.

General coverage:

FPW75 *Nuclear Tracks in Solids*, R. L. Fleischer, P. B. Price, and R. M. Walker, Univ. of California Press, Berkeley, 1975, 605 pages.

DB87 *Solid State Nuclear Track Detection*, S. A. Durrani and R. K. Bull, Pergamon Press, Oxford, 1987, 300 pages.

Track fundamentals, track etching, and miscellaneous applications:

S90 *Ion Tracks and Microtechnology*, R. Spohr, Vieweg, Braunschweig, 1990, 272 pages.

Radon techniques and applications:

ID97 *Radon Measurements by Etched Track Detectors*, R. Ilić and S. A. Durrani, editors, World Scientific, Singapore, 1997, 387 pages.

Track dating:

WH92 *Fission-Track Dating*, G. A. Wagner and P. van den Haute, Enke, Stuttgart, 1992, 285 pages.

In addition to the above, characteristics of etched track detectors are documented in a vast general literature, in the journal variously called *Nuclear Track Detection* (1977–78), *Nuclear Tracks* (1979–83), *Nuclear Tracks and Radiation Measurements* (1984–1993), and *Radiation Measurements* (1994 to the present), and in what are now every-two-year conference proceedings on nuclear tracks in solids, the first of which was in 1969.

SUBJECT INDEX

AUTHOR INDEX

The beginnings and endings of all human undertakings are untidy, the building of a house, the writing of a novel, the demolition of a bridge, and, eminently, the finish of a voyage.

John Galsworthy in Over the River, *Chapter I*

MIX
Papier aus verantwortungsvollen Quellen
Paper from responsible sources
FSC® C105338

If you have any concerns about our products,
you can contact us on
ProductSafety@springernature.com

In case Publisher is established outside the EU,
the EU authorized representative is:
Springer Nature Customer Service Center GmbH
Europaplatz 3, 69115 Heidelberg, Germany

Printed by Libri Plureos GmbH
in Hamburg, Germany